THE ENVIRONMENT AND ITS RESOURCES

THE ENVIRONMENT AND ITS RESOURCES

RAUL A. DEJU

Senior Author
Gulf Research & Development Company and (part-time)
University of Pittsburgh,
Pittsburgh, Pa.

ROSHAN B. BHAPPU

New Mexico Institute of Mining and Technology
Socorro, New Mexico

GEORGE C. EVANS

U.S. Department of the Interior
Bureau of Mines and (part-time)
University of Pittsburgh,
Pittsburgh, Pa.

ARMANDO P. BAEZ

Instituto de Geofísica
Universidad Nacional Autónoma de México
México, D.F.

GORDON AND BREACH, SCIENCE PUBLISHERS
NEW YORK LONDON PARIS

GORDON AND BREACH, Science Publishers, Inc.
440 Park Avenue South
New York, N.Y. 10016

Editorial office for Great Britain :
Gordon and Breach, Science Publishers Ltd.
41-42 William IV Street
London W.C.2, England

Editorial Office for France
GORDON & BREACH
7–9 rue Emile Dubois
Paris 14e

Library of Congress Catalog Number **72–78467**

ISBN 0 677 14120 3 (cloth).

Printed in the United States of America

PREFACE

This book has been in the making for the past four years. The initial idea for the book came about while the senior author was at New Mexico Institute of Mining and Technology in 1968. At that time it was decided that the purpose of the book should be to present a unified, well-coordinated picture of the physical environment and its resources. Four years later we hope to have achieved this purpose.

The book is intended both as a source-book for managers, scientists, engineers, and government officials and as a text to train students in the various facets of environmental science. While the book is written in reasonably simple language, it contains a tremendous amount of information and hundreds of references for those desiring additional details. For instructors desiring to use this book as a text, the book offers an appendix containing over one hundred problems. All the problems included have been used by the authors at the freshman-sophomore level. The book can easily be covered by a two-semester course at this level.

After a brief introductory chapter, the book discusses in some detail the four fundamental parts of the environment, namely, land forms, continental water, oceans, and air. These can be classed as the "common property resources" of the earth. In addition, natural and man-made resources easily subject to private ownership, such as, minerals, fuels, and wastes that can be re-used are treated in additional chapters. The effect that waste disposal has on the environment is also considered. The role of man as he affects all areas of environmental science permeates the entire book.

Finally, we should note that we have tried to keep the discussion as unbiased as possible by exposing all possible sides of a problem before stating any conclusions. It is our hope that this book will not only serve to inform its readers but will also serve to inspire them to take a more responsible attitude towards our environment and its resources.

Raul A. Deju
Pittsburgh, Pa.
Roshan B. Bhappu
Socorro, New Mexico
George C. Evans
Pittsburgh, Pa.
Armando P. Baez
Mexico, D.F.

NOTE FROM THE AUTHORS AND THE PUBLISHER:

Dr. Deju contributed all of his material for this book while he was Chairman, Department of Hydrology, Institute of Geophysics, University of Mexico. Since he wrote this material he has joined Gulf Research & Development Company and (part-time) The University of Pittsburgh, in Pittsburgh, Pennsylvania.

Dr. Bhappu is presently Acting Chairman, Department of Metallurgy and Materials Engineering and Vice Pres. for Research N.M.T.R.F., New Mex. Institute of Mining and Technology. He is also associated in a part-time capacity with the State Bureau of Mines and Mineral Resources, Socorro, New Mexico.

Mr. Evans is presently Geologist, U.S. Department of the Interior Bureau of Mines, Eastern Field Operations Center and (part-time) Lecturer at The University of Pittsburgh, Pittsburgh, Pennsylvania.

Mr. Baez was formerly the head of the Airglow Research Laboratory at the Institute of Geophysics, University of Mexico. He is currently head of the Water Pollution and Hydrological Laboratory of the same institution.

Drs. Deju and Bhappu and Mr. Evans worked at New Mexico Institute of Mining and Technology prior to 1969. Dr. Deju and Mr. Baez worked together at the Institute of Geophysics of the University of Mexico until Dr. Deju joined Gulf Research & Development Company. The book is the result of the mutual interaction of the four authors.

OPINIONS PRESENTED IN THIS BOOK ARE THOSE OF THE AUTHORS AND DO NOT NECESSARILY REPRESENT THE OPINIONS OF THE PUBLISHER OR THOSE OF THE AUTHORS' PRESENT OR PAST EMPLOYERS.

ACKNOWLEDGMENTS

Many individuals and agencies have graciously collaborated with the authors by reviewing the manuscript or contributing material to it. Special thanks are due Drs. Marvin H. Wilkening and Geoffrey Purcell of New Mexico Institute of Mining and Technology, Mr. L. P. McDonnell of Dorr-Oliver, Inc., Mr. George Crouter (photographer) of Empire Magazine of The Denver Post, Mr. Ralph S. Blois of Leupold & Stevens, Inc., Mr. Paul B. DuMontelle of the Illinois State Geological Survey, Mr. Raul Belmont of the Instituto de Geofísica at the University of Mexico, Mr. John C. Stormont (photographer) of Moyer-Stormont Productions Undersea Media, and Mr. R. M. Campau of Ford Motor Company.

The assistance of New Mexico Institute of Mining and Technology personnel and personnel from the State Bureau of Mines and Mineral Resources in Socorro, New Mexico is greatly appreciated. Pictorial and tabular contributions were received from United States Geological Survey, United States Bureau of Mines., International Underwater Explorers Society, Deepsea Ventures, Inc., the United States Naval Oceanographic Office, and the Department of Health, Education and Welfare. Finally, thanks are extended to the many individuals that in one way or another helped in the preparation of this book.

TABLE OF CONTENTS

CHAPTER 1. INTRODUCTION

CHAPTER 2. LAND FORMS OF THE ENVIRONMENT

CHAPTER 3. WATER RESOURCES

CHAPTER 4. THE OCEANS

CHAPTER 5. MINERAL RESOURCES

CHAPTER 6. SOLID WASTE DISPOSAL AND REUSE

CHAPTER 7. THE ATMOSPHERE

Chapter 1

INTRODUCTION

1.1 *MAN AND THE ENVIRONMENT*

In the past few years man's awareness of the environment has greatly increased, mostly, as a result of the cries of ecologists and as a result of observing the environmental damage already done.

The environment is everything that surrounds us. We are also part of the environment. As such, the environment is a huge pool of resources that helps us satisfy our needs and fulfill our wants. These resources can be grouped into four categories, namely,

a. the earth's landscape,
b. waterways,
c. air, and
d. mineral matter (including oil, water and gas).

The last category covering mineral resources, etc. is economically speaking different from the first three. Minerals have to be classed in a realm subject to private ownership while the other three categories are more properly defined as "common property resources." Kneese[1] defines such resources as a group that "encompasses those valuable attributes of the natural world which cannot be, or can be only imperfectly, reduced to individual ownership, and therefore, do not enter into the processes of market exchange and the price system." Such common property resources cannot be managed on the basis of the price system. In addition, none of the environmental resources is infinitely abundant. Therein, lies the trap into which we have fallen and which has led to our widespread pollution. If man is to survive more than just a few generations we must re-examine the man-earth relationship and attempt to understand, manage, and control, as adequately as possible, all our environmental resources. In addition, we must minimize all types of pollution and avoid waste. Stringent national and international regulations must be enforced in all the above areas.

In our reevaluation of the interrelation between man and the environment one must not forget the dynamic nature of the environment. For instance, 11,000–12,000 years ago glaciers covered the area occupied today by the city of Toronto (Canada), 500 years ago the north-american continent was in a wild, undeveloped state, with very few men roaming around. We cannot expect to return the environment to such conditions. Nor can we attempt to preserve (as some pseudo-ecologists try to make us believe) the world in a state of pristine, immutable beauty. What man can and should do is recognize himself as an important geologic agent and attempt to be aware of the environmental side effects to which each of his decisions leads. To accomplish this we must educate large masses of our population, and make them aware of the importance of the environment and its resources.

Such a study of the environment is a very interdisciplinary task. In this book we are dealing only with those aspects of the environment that fall within the realm of natural science. In addition, we will look at the waste and pollution problems imposed on us due to our mismanagement of the environment. The health hazards posed by these problems will also be treated.

There have been numerous books written about the environment in the past few years.[2–4] However, many of these are either a collection of widely diverse essays or attempts to scare the reader with grossly exaggerated doomsday predictions. There are also some treatises of a more specialized nature covering specific aspects of the environment such as, water,[5,6] wastes,[7] and environmental law.[8] In addition, numerous articles are available that give an idea of the natural science aspects of the environment and what programs can be instituted to better inform the public about them.[9–11]

At the root of the problem lies "environmental pollution and waste." By pollution we will mean in this book any undesirable degradation of the environment be it natural or man-made. By waste we mean any undesirable substance accumulated as a result of our inefficient utilization of the environment's resources. These two concepts, "pollution" and "waste," are basic to all the ideas to be discussed later in this text. One of the fundamental steps that needs to be taken in order to solve the environmental crisis in which we find ourselves is to minimize both waste and pollution.

There are various forms of pollution, such as that of air,[12] water,[6] and land[7] (garbage disposal, geomorphological degradation, etc.). In the future, other problems such as noise pollution[13] may become very important. Action must be taken now to solve not only our present problems but to anticipate and prevent future pollution difficulties.

A most serious strain on the environment that greatly increases our pollution problems is population growth. Presently, the human population stands at roughly 3.7 billion and is expected to double within the next 35 years. This is quite a change from 2000 years ago when only 150 million people inhabited the entire earth (less than the population of the United States today[7]). It does not take an expert in demography to see the dangers of this overpopulation. The noted biochemist Isaac Asimov[14] and many others have extensively discussed this subject. In addition, it should be noted that the optimum population for the planet earth is certainly much less than the maximum possible population. The present rapid population growth in the earth as a whole cannot remain unchecked. One of two things must happen. The first alternative is to increase the usage of birth control in countries where the population growth rate exceeds certain allowable limits. The second alternative is to raise the death rate. This will automatically happen if the first alternative is not adopted and will be the result of widespread famine and pestilence. Which alternative to take is for man to decide. However, such a decision must be taken now and not further delayed.

The subject of the interrelation of human population growth and the environment has just been mentioned in passing and will not be discussed further in the book. This topic is best left for a textbook on human ecology. This superficial coverage, however, should not diminish in the mind of the reader the importance of the subject.

Mankind has basically two choices, either to keep the planet healthy or die with it.[4] If we continue to operate (as we have done in the past) on the *principle of superabundance of resources* the last alternative will be inevitably accelerated. However, if we apply our technology and management techniques wisely to our environmental resources we can assure that many coming generations will enjoy the fruits of our wisdom. Planning is the key to our problem, and such planning must be started now.

1.2 PLANNING THE UTILIZATION OF ENVIRONMENTAL RESOURCES

In planning the optimal utilization of our environmental resources many factors need to be taken into account. First, we must accurately analyze the resources needed to satisfy present demands. Each country has to closely examine how to utilize their available resources to satisfy the needs of their people to a maximum. They must also examine which resources to acquire from other countries to complement certain needs and at the same time determine which resources they have in excess so that they can be exported in an attempt to have a favorable national trade balance.

Such a picture of present resource exploitation and commerce should be coupled with an accurate evaluation of future demands and reserve estimations. One must weigh the advantages and disadvantages of depleting a resource so rapidly that it would all be gone before a suitable substitute is found. In short, a serious scrutiny of the supply-demand picture in both a short and a long-range scale is needed on a national level before going all out in the exploitation of a given resource. Exploitation must be geared to maximize the satisfaction of human demands now and in the future.

In addition to planning the exploitation of our environmental resources, we have a second, most-difficult task ahead of us. Priorities must be set and decisions must be made as to what type of an environment we want for us and future generations. Such priorities must be set prior to and concomitant with the exploitation scheme discussed above.

Finally over the coming years we must strengthen our knowledge of the physical, chemical, and biological world in order to try to optimize utilization of known resources and spur the discovery of new ones. Such research is particularly important in optimizing the use of minerals, fuels, and food.

This book deals with the scientific aspects of the utilization of environmental resources. Therefore, planning the use of these resources is one of our prime subjects for discussion. Such planning can be divided into six areas, each of which is treated in a separate chapter of the book. These areas are:

1. the land surface,
2. water resources,
3. ocean resources,
4. minerals and fuels,
5. the atmosphere, and
6. wastes.

The land surface is a prime area of concern to environmentalists. The study of the morphology of the land, its deterioration, and change due to natural and/ or man-made causes is the subject of *environmental geomorphology*. One of the most important goals of this discipline is the development of a concept of optimal utilization of the land surface, be it for agriculture, recreation, urban development, or other uses. Such a concept in the future will most likely include the idea of multiple sequential land use, that is, a given area may be used for recreation first, later on for resource extraction, and perhaps later on for urban expansion. In addition, the effects of man and other geologic agents on the land's morphology must be included at the beginning of our planning scheme in the area of environmental geomorphology.

Water resources are basic for mankind's survival. The need for water has been intensified due to industrial development, the increase in the use of household appliances, and the application of intensive farming techniques. At the same time, the advantages brought about by our technological revolution are responsible for polluting many of our sources of water. A planning scheme in the area of water resources should attempt to optimize water usage and minimize pollution. In general terms, it is probable that the entire concept of water supply will undergo drastic reshaping within the coming decade. We must move into the integration of a regional supply concept which is the subject of the emerging discipline of *regional hydrology*.[6]

The oceans are a most important source of many varied resources. In thinking of the ocean resources, one commonly thinks of the fishing industry, however, many other resources from the ocean are presently being exploited.[15] For instance, present exploitation of offshore oil accounts for 1/6th of the world's total production of this resource. The production of gem-quality diamonds in southwest Africa has just begun and is already producing gems worth over four million dollars per year. Many other resources will be extracted from the oceans in the future as technology develops, and as the economics of such ventures become more favorable. In planning for such future developments one of the most important areas will be the political issue as to who will control the ocean. This crucial problem must be solved by international agreement between all nations concerned. In addition to this political issue, we must reduce and control the pollution of the oceans if we are to preserve them as pools of food for human consumption.

Planning in the area of mineral resources and fuels is of critical importance to an industrialized society. Such planning must include the technical areas of exploration, production, and transportation, as well as the economic areas of supply, demand, marketing, and transportation economics. All such planning should be tied-up in order to attain a comprehensive *national minerals policy*. As in all the other areas being considered such policy must include both short and long-range planning.

One of the most critical problems facing scientists today is what we can term the *energy-environment problem*. This problem is two-fold. First, it involves the degradation of the environment as a result of producing energy, and secondly, it involves the need for developing new energy sources to satisfy increasing demands. Much research is presently underway to develop new sources of energy. Such research involves work in developing the fast-breeder reactor, development of geothermal power, coal gasification research, etc. A national fuels policy as part of an integral national minerals policy should foster such important research activities.

One part of the environment that is not usually thought of as a resource is the atmosphere of our planet. Nonetheless, we could not live very long if we did not have air to breathe. The atmosphere is a "common property resource" that we must learn to manage if our cities are to survive. A comprehensive plan to abate air pollution should have top priority as part of governmental budgets.[16] Such a plan may require drastic changes in the nature of our urban areas and in the modes of transportation that we commonly use.

Up to now we have depended for waste disposal upon diluting our gaseous wastes into the atmosphere and our liquid wastes into our waterways. Our solid wastes have mostly been dumped onto the land as trash, many remaining in an unsightly, unsanitary condition. About 200 billion tons of trash are dumped every year in the United States alone. By 1985 this number should double if present trends continue. A comprehensive national plan in the area of waste disposal should:

1. consider the possibility of improving present waste treatment and disposal methods;
2. encourage recycling of materials and minimization of waste; and
3. impose stringent regulations on waste disposal.

Research in all these areas is also of the utmost importance.[7,17]

One final area needs to be considered in planning a national environmental policy, namely, the effects of pollution on our health. Our environmental planning should gear our technological development in such a way that our health is in no way impaired by pollution, and so that we are guaranteed adequate food supplies. The effects of air pollution on human health are particularly important and are included in our discussion of the atmosphere in Chapter 7.

In planning an overall scheme for the optimal utilization of our environmental resources one must devise each of the individual plans discussed above in such a way that they complement each other and offer as a result:

1. an integral improvement in the quality of life;
2. a preservation of the environment and its resources in a way satisfactory to us and hopefully to future generations; and
3. the possibility of advancing the development of our society.

1.3 ADMINISTERING THE ENVIRONMENT AND ITS RESOURCES

As we discussed in the preceding section planning for the optimal utilization of our environmental resources is a most difficult task. However, if progress is to continue we must tackle such a problem now.

In administering our environmental resources and in carrying out a

TABLE 1.1.

Role of Government, Industry, and the Individual in the Administration of Environmental Resources and Antipollution Activities.

1. *Federal Government*
   a. Set-up policies and national goals.
   b. Serve as a source of information.
   c. Support environmental research and antipollution activities.
   d. Control pollution by:
      i. establishing guidelines,
      ii. policing the compliance with such guidelines, and
      iii. penalizing those who do not comply.
   e. Help local and state governments by setting up a revenue sharing system to support environmental projects.
   f. Provide incentives to those who comply with pollution guidelines and those whose standards are higher than the minimum required by law.
2. *State and Local Government*
   a. Sponsor projects designed to advance the quality of life available to the residents of the state or the community.
   b. Complement the role of the Federal Government in all the areas listed above.
3. *Industry*
   a. Take a responsible attitude toward the management of environmental resources and antipollution activities.
   b. Support research aimed at improving product performance, foster recycling, increase plant efficiency, and reduce to a minimum all types of pollution due to plant operations.
   c. Comply with government guidelines.
   d. Cooperate, inasmuch as possible, with government units in solving the environmental problems.
   e. Support university programs aimed at improving the environment.
4. *The Individual*
   a. Comply with antipollution regulations.
   b. Cooperate with government units in solving environmental problems.
   c. Learn to be more careful, less destructive, and less wasteful in our daily activities.

battle against pollution, government units, industry, and individual citizens must all play a role. Table 1.1 lists some of the activities that each of these parties should be responsible for. This table is not an exhaustive list because such a list would require all the pages of this book and more. Table 1.1 is simply intended to give the reader an idea of some of the basic tasks that must be carried out. Universities can also play an important role by carrying out research and training students in the environmental sciences.

Figure 1.1. Trash on the Platte River near Denver, Colorado. Industry alone is not responsible for pollution, man himself helps a great deal! (*courtesy The Denver Post*).

The most helpful step that everyone can take in managing our resources and solving the problem of environmental pollution is to learn to be more responsible, more careful, less destructive, and less wasteful in our daily activities. Much will be accomplished by taking such a simple step. Situations such as individuals dumping their trash in rivers (see figure 1.1) are quite inexcusable examples of the actions of irresponsible citizens. Planning is the key to managing the environment and its resources, and *responsibility* is the key to such planning.

1.4 REFERENCES

1. Kneese, Allen V. (1970) *Protecting Our Environment and Natural Resources in the 1970's*, a prepared statement for "The Environmental Decade" (Action Proposals for the 1970's), Hearings before a Subcommittee of the Committee on Government Operations, House of Representatives, 91 Congress, 2 sess. (Washington, U.S. Government Printing Office), reprinted by Resources for the Future, Inc., Washington, D.C., pp. 190–197.

2. Jarrett, Henry (ed) (1958) "Perspectives on Conservation—Essays on America's Natural Resources," The Johns Hopkins Press, Baltimore, Md., 258 pp.
3. American Chemical Society (1969) "Cleaning Our Environment, The Chemical Basis for Action", American Chemical Society, Washington, D.C., 249 pp.
4. Rose, J. (ed) (1969) "Technological Injury", Gordon and Breach, Science Publishers, Ltd., London, England, 224 pp.
5. DeWiest, R. J. M. (1965) "Geohydrology", John Wiley & Sons, New York, N.Y., 366 pp.
6. Deju, R. A. (1971) "Regional Hydrology Fundamentals", Gordon and Breach, Science Publishers, Inc., New York, N.Y., 204 pp.
7. Marx, Wesley (1971) "Man and his Environment: Waste", Harper & Row, Publishers, New York, N.Y., 179 pp.
8. Murphy, Earl Finbar (1971) "Man and his Environment: Law", Harper & Row, Publishers, New York, N.Y., 168 pp.
9. Meyer, W. G. (1971) *AAPG—Environmental Geology Program*, Bulletin Am. Assoc. of Pet. Geol., volume 55, pp. 175–176.
10. Frye, John C. (1967) "Geological Information for Managing the Environment", Illinois State Geological Survey, Environmental Geology Notes #18, 12 pp.
11. ——— (1971) "A Geologist Views the Environment", Illinois State Geological Survey, Environmental Geology Notes #42, 9 pp.
12. Middleton, John T. (1969) *Air Pollution: Where are We Going*?, paper presented at a conference on environmental pollution sponsored by the Junior League of Los Angeles and the Rand Corporation, Los Angeles, California, December 6, 1969, repr. 1970, U.S. Government Printing Office, 8 pp.
13. Anonymous (1970) *Noise-Fourth Form of Pollution*, Environmental Science and Technology, volume 4, pp. 720–722.
14. Asimov, Isaac (1970) *The Case Against Man*, The Pittsburgh Press Sunday Roto, 8/23/70, pp. 28–33.
15. Morello, T. (1970) *Fight Over Ocean Wealth*, The Pittsburgh Press Sunday Roto, 12/13/70, pp. 46–49.
16. Anonymous (1970) *Federal Budget Stresses Quality of Life*, Environmental Science and Technology, volume 4, pp. 193–194.
17. Sheffer, H. W., Baker, E. C., and Evans, G. C. (1971) "Case Studies of Municipal Waste Disposal Systems", United States Department of the Interior, Bureau of Mines Information Circular 8498, U.S. Government Printing Office, Washington, D.C., 36 pp.

Chapter 2

LAND FORMS OF THE ENVIRONMENT

2.1 *INTRODUCTION*

Land forms are from a layman's point of view the most obvious expression of nature. We see mountains, valleys, shorelines, river banks, etc. as we travel from one place to another. For those of us fortunate enough to live in a rural or suburban environment these expressions of nature surround us. In our study of the environment land forms are a logical place to start. The science that is concerned with such an analysis is called "geomorphology." The term geomorphology comes from the combination of the roots "geo", "morphe", and "logos" and etimologically signifies "the study of land forms."

Geomorphology is at times included as a part of physical geography while at other times it is treated as part of geology.[1] Some authors have used the term geomorphology as being synonymous with physiography. This is incorrect since physiography is a much more general term than geomorphology and includes the description of natural phenomena in general.

We will define geomorphology as that phase of environmental science dealing with the study of:

(a) The form and shape of the earth;
(b) The dynamic processes that build and destroy land forms;
(c) The configuration of surface and subsurface land forms; and
(d) The distribution of land and water masses.

A geomorphologic analysis of the environment that surrounds us is important from four points of view:

(a) It helps us understand the natural and man-made processes that shape our environment and make it look the way it does;
(b) It gives us a greater awareness of nature and land forms;
(c) It gives us an idea as to how man has affected the natural environment; and
(d) It furnishes us with tools applicable in many other facets of environmental and resource science.

Geomorphologic studies are basically of two types: descriptive and quantitative. Descriptive geomorphology is concerned with explaining the occurrence of land forms by the use of visual observation. Much of the early work in geomorphology was of this type and the descriptions of the observations were usually embellished by the use of very elaborate language. A good example of one such description is Playfair's Law[2] which was John Playfair's (1748–1819)[3] way of describing how tributaries and their valleys enter into the mainstreams and their valleys. The law states:

> "Every river appears to consist of a main trunk, fed from a variety of branches, each running in a valley proportioned to its size, and all of them together forming a system of vallies, communicating with one another, and having such a nice adjustment of their declivities that none of them join the principal valley either in too high or too low a level."

Starting with the second quarter of this century, however, descriptive geomorphology has given way to quantitatively expressing the laws that describe the processes that in one way or another cause changes in the earth's features.[4] Quantitative geomorphology is exemplified by the hydrophysical approach to the study of the erosional development of streams and their drainage basins which was pioneered by Horton.[5] In addition, the work of geomorphologists has recently been aided by the use of photographs and data obtained as a result of the space program.

One thought which is most important in understanding land forms is their *dynamic nature*. Our environment is constantly changing not only as a result of the work of man, but also as a result of the work of natural processes. John C. Frye of the Illinois State Geological Survey has described this characteristic of the environment very beautifully as follows:[6]

> "The environment is a dynamic system that must be understood and accomodated by man's activities, rather than a static, unchanging system that can be 'preserved'. The living or biological systems of the earth are generally understood to be progressively changing, but much less well understood by the public is that the nonliving, physical aspects of the earth also undergo change at an equal or greater rate.
>
> Clearly, a dynamic system is more difficult to understand fully, and it is more difficult to adapt man's activities to a constantly changing situation than to an unchanging or static system. On the other hand, the very fact of constant change opens many avenues of modification and accomodation that would not be available in a forever constant and unchanging system."

To describe the geomorphologic processes that constantly reshape the land forms of the earth requires many more pages than we can devote to this subject in this book. Our aim here will only be to illustrate the effect of various geomorphologic processes in regulating the nature of land forms. In addition, we will consider the geologic hazards resulting from geomorphologic processes and the application of remote sensing techniques to improve our understanding of land forms and geomorphologic processes.

2.2 HISTORICAL DEVELOPMENT

The study of land forms has had a long history and can be traced back to 450 B.C. and the time of Herodotus, Plato, Aristotle, and others. Among the accomplishments of the Greeks one can note that they crudely understood the process of sedimentation; they noted that sea levels had changed throughout geologic time; they noted that rivers were responsible for erosion and deposition; and were able to construct a very rudimentary picture of the hydrologic cycle. On the minus side, however, we must note that the Greek philosophers firmly believed that earthquakes were an expression of the anger of the Gods.

Developments by Roman thinkers in the area of geomorphology were also as impressive. Strabo, for instance, was able to recognize the occurrence of subsidence and uplift. However, just as the Greeks, the Roman thinkers attributed to the Gods any phenomenon they could not logically explain.

Important accomplishments in the description of land forms took place in Arabia during the X century. A somewhat forgotten, but most important work by unknown authors entitled "The Discourses of the Brothers of Purity" was written between 941 and 982 A.D.[7] The Discourses appear in four volumes and contain among other things:[8]

(a) a perfectly acceptable description of the metamorphic cycle, and
(b) the first accurate description of peneplanation, pond evolution, epicontinental seas, weathering, erosion, and transport by streams and wind.

The Discourses presented these ideas in simple but exacting language. For instance, they described the continents[8] as eggs that are immersed in water, thus paving the way for the important concept of isostasy.

In addition to the Brothers of Purity, Arab thinkers like Avicenna and Avirros contributed much in the area of geomorphology. Unfortunately, their thoughts did not reach European scholars of this period. In fact, almost no progress took place in Europe during the Dark Ages.

European scholars of the late XV century used to invoke the "Principle of Catastrophism" whenever they were interested in describing the origin of some land form. According to Catastrophism land forms were either created by God as such or were the result of an *instantaneous* cataclysm. Thus, European scholars of the Middle Ages did not recognize the occurrence of slow geologic events.

During the XVII century Steno developed the "Principle of Original Horizontality" and the "Law of Superposition"[9]. The first of these laws said that sedimentation underwater produced strata that were initially parallel to the surface on which they were laid. The Law of Superposition ascertained that in an undisturbed sedimentary sequence the most recent sediments are at the top and the oldest at the bottom.

One of the problems during the XVII and XVIII centuries in Europe was the fear of heresy. For instance, the Frenchman Buffon suggested that the creation of the earth took longer than six days. Such a thought was considered heretical and he was asked to retract.

Other important developments during the XVIII century in Europe include the work of Hutton who in 1785 postulated "The Principle of Uniformitarianism".[10] This is the now famous idea that "the present is the key to the past." This principle was later popularized by John Playfair and Sir Charles Lyell.

One of the prominent geomorphologists of the late XIX century was W. M. Davis.[11–13] He believed that all land forms go through a very orderly sequential development. This line of thought led him to enunciate his controversial concept of the geomorphologic cycle (sometimes termed geographical cycle). This concept indicated that as a result of erosive effects a land form would pass through stages of youth, maturity, and old age and would eventually lead to a featureless condition termed *peneplain*.[14] Walther Penck[15,16] did not believe in the geomorphologic cycle of Davis but admitted the possibility that land forms need not go through all developmental stages, nor need to end up in a peneplain condition. Penck is also famous for having postulated the concept of *slope retreat* which relates slopes to changes in uplift and erosion rates.

The idea that land forms are in a dynamic state of evolution caused by energy changes was not widely accepted until very recently, and still some geologists are in disagreement. Hack in 1960 proposed his theory of "dynamic equilibrium".[17] This principle provides a more logical basis for the interpretation of topographic forms in an erosionally graded landscape. Hack says[17] that every channel and slope in a given erosional system must necessarily be adjusted to every other channel and slope. He explains differences in form and relief in terms of spatial relations rather than in terms of sequential evolutionary development through time. Hack's main contribution is the idea that topographic forms evolve as a function of energy changes.

Ever since man started observing nature he has wondered about the origin of land forms. With the recent advent of satellite photography and remote sensing techniques, his observational scope has been greatly widened. These improvements in observational techniques coupled with new developments in quantitative geomorphology are mostly responsible for the progress achieved in the last few years in understanding land forms.

2.3 GEOMORPHOLOGIC PROCESSES

By geomorphologic processes we understand all the physical and chemical changes that control the nature and appearance of land forms. Such processes are classified into two categories,[4] namely,

(a) exogen(et)ic processes which are the ones that act from outside the earth's surface, and
(b) endogen(et)ic processes which act from inside the earth's crust.

TABLE 2.1.

Geomorphologic Processes

Exogenetic Processes

1. Weathering
   i. mechanical disintegration
   ii. chemical action
2. Mass wasting (slides and subsidence)
3. Stream erosion
4. Erosion by ocean waves
5. Turbidity currents
6. Glacial erosion
7. Wind erosion
8. Groundwater dissolution
9. Denudation and accumulation as a result of the work of animals (including *man*) and plants etc.

Endogenetic Processes

1. Volcanism
2. Seismic events
3. Diastrophism
   i. orogenic or mountain-building processes
   ii. epeirogenic (regional uplift) processes etc.

Table 2.1 shows an outline of some of the most important geomorphologic processes. Many of these processes such as the effect of waves and wind have been termed to be *gradational* by many authors,[18] that is, they tend to produce an equally-leveled landscape. In general, we can classify all gradational processes as being either *denudation* or *accumulation* processes. The former are those that tend to reduce the level of existing land forms while the latter tend to raise this level. Many geologists have used other terms such as degradation as synonymous of denudation and planation as synonymous of gradation.

Among the exogenetic processes, three of them are of utmost importance, namely, river erosion, weathering, and the work of man. The last one will be treated in section 2.7 and thus we will restrict our discussion here to the other two.

River erosion has a double effect on the land environment. First, it removes rock from the river's bed and banks, and second it deposits this rock load on other parts of the river's course and tends to form a delta. A delta is a flat depositional plain built by the river as it enters a body of quiet water such as the ocean or a lake. River erosion therefore encompasses the following processes:

(a) rock reduction,
(b) material transport,
(c) cavitation,
(d) abrasion,
(e) rock accumulation along bed and banks, and
(f) delta formation.

The erosive capacity of a river can be evaluated using as an index the maximum size of particle of a given type it can carry under the existing flow regime.[4] Another quantity one needs to calculate the erosive capacity of a river is the total particle load it can carry.

The erosive capacity of a river greatly increases during storms and decreases in times of drought. Important changes in the morphology of a basin can occur as the result of intense storms.

A complete analysis of the erosive capacity of a river must also include a study of the process of rock accumulation. The removal and accumulation processes do not occur independently but go side by side. For a more complete description of the geomorphological implications of river erosion the reader should refer to more advanced works.[19,20]

The process of weathering can be defined as the result of the action of the elements of climate in altering the color, texture, and composition of surface rocks. Such altered rock products are said to be "weathered." There are two types of weathering, namely, mechanical and chemical. As will be discussed in section 2.5, climate plays a most important role in regulating both of these processes, especially chemical weathering. If there were no chemical weathering on our planet our landscape would look much like the moon.

Five processes can be included within the definition of chemical weathering. These are:

(a) hydration (adsorption of water), e.g. the conversion of hematite to limonite;
(b) oxidation, e.g. the oxidation of olivine;
(c) hydrolysis (formation of hydroxyl), e.g. hydrolysis of feldspars;
(d) carbonation (attack by CO_2), e.g. when a silicate reacts with water and carbon dioxide the metal cation from the silicate structure will react with the carbon dioxide to form a metal-carbonate that will be carried in solution; and
(e) dissolution, e.g. dissolution of calcite to form calcium bicarbonate.

Recent studies on weathering of minerals such as the silicates[21] have pointed out the importance of adsorption-desorption processes and of mineral structure. All adsorption phenomena taking place at the rock-water interface during weathering can be conceived as adjustments to produce a "minimum of free energy difference between the crystal, the surface, and the external environment."[22]

Mechanical weathering involves the mechanical disintegration or breakdown of rock. This can be accomplished as the result of frost action (wedging), crystal growth, mass wasting, plant or animal action, day-to-day temperature changes, sandblasting, etc.

Weathering is of utmost importance to soil mechanics since much of what we call soil is the result of the action of weathering. Basically, soil profiles can be said to be the result of *rock rotting* in a very literal sense. For a more detailed look at the formation of soils by weathering the reader can refer to the excellent work by Jenny.[23]

Among the endogenetic processes one can include diastrophism and volcanism. Diastrophism includes all those geologic processes that lead to deformation of the earth's crust and the production of mountains, continents, orogenic belts, and ocean basins. Volcanism refers to the action of volcanoes, namely the issuing of steam, gases, and/ or lava out of vents in the earth's crust.[24]

All in all the geomorphologic processes that shape our landscape are extremely varied. These processes have been active in the past and are active now. The environment is dynamic and continually subject to change.

2.4 GEMORPHOLOGIC PROCESSES AND TOPOGRAPHY

The land forms that we see today can be said to be the result of the interplay of the various geomorphologic processes that were described in the previous section. Taking a very simplistic attitude we can say that endogenetic processes for the most part tend to build the relief that was destroyed by exogenetic processes.

Each geomorphologic process is characterized by leaving certain distinctive topographic marks, thus accounting for the variability of land forms that we can observe. Table 2.2 illustrates some of the characteristic features resulting from the action of winds and desert erosion, igneous activity, glaciers, streams, groundwater, and the work of man.

TABLE 2.2.

Topographic Features Characteristic of Various Geomorphologic Processes.

1. Desert erosion and wind action produce characteristic features such as:
   (a) steep slopes of bare rock
   (b) bahada slopes
   (c) playa floors
   (d) pediments
   (e) sand dunes

2. Streams produce characteristic features such as:
   (a) floodplains
   (b) peneplains
   (c) deltas
   (d) alluvial fans
   (e) dendritic patterns
   (f) stream terraces
   (g) trellis patterns

3. Glaciers produce characteristic features shch as:
   (a) cirques
   (b) smoothed rocks
   (c) moraines
   (d) inmature drainage
   (e) drumlins
   (f) U-shaped valleys
   (g) bergschrund
   (h) hanging valleys

4. Igneous activity produces characteristic features such as:
   (a) composite volcanoes
   (b) dikes
   (c) laccoliths
   (d) lopoliths
   (e) pillow lavas
   (f) plateau basalt flows
   (g) plugs
   (h) plutons
   (i) shield volcanoes
   (j) sills
   (k) spines
   (l) volcanic mudflows

5. Groundwater produces characteristic features such as:
   (a) sinkholes
   (b) cemented sandstone beds
   (c) dripstone
   (d) solution channels
   (e) caverns
   (f) Karst topography

6. Man produces characteristic features such as:
   (a) roadways
   (b) urban areas
   (c) waste dumps
   (d) abandoned mine holes

Most of the topographic features that we observe today are very young. Ashley[25] estimated that at least 90% of our present topography is post-Tertiary (2 million years or less) and possibly as much as 99% of the world's topographic features are post-Miocene (25 million years or less). There is abundant information from all over the world to back these statements. For instance, the Pacific Coast of the United States "is notable for changes that are certainly not older than the Quaternary (1 million years or less) and many of which appear to date from the last few thousand years."[25]

Most of the world's highest mountains seem to have reached their present elevation about the end of the Tertiary (1–2 million years ago) and a great deal of their present shape is the result of ice cutting during Pleistocene time.

The various topographic features that result from all the different geomorphologic processes in turn affect many subsequent geologic changes. For instance, topographic relief is a most important controlling factor of weathering.[26] It affects both the rate of weathering as well as the nature of weathering products because topographic relief controls:

(a) rates of surface runoff,
(b) rates of subsurface drainage, and
(c) erosion rates.

Loughnan[26] in his recent book summarizes the effect of topography on chemical weathering as follows:

> "The ideal conditions for chemical weathering are attained on rolling to gently sloping uplands where surface runoff is not excessive and the subsurface drainage is unimpeded. Under such conditions, the weathered zone may extend to a depth of 100 feet or more. However, even here local variations in topography create distinctly different environments which may find expression in contrasting weathered products."

2.5 GEOMORPHOLOGIC PROCESSES AND CLIMATE

Climate is one of the controlling factors of the earth's landscape. Land forms have been greatly affected not only by present climatic conditions but also by past climates, especially, Pleistocene glaciation.

Climate variations account to a large extent for the differences in rates and kind of geomorphologic processes that take place in a given area. In humid tropical regions weathering is a maximum and rocks undergo drastic chemical decay. The abundance of vegetation in such areas tends to inhibit lateral erosion while permitting runoff. In humid tropical areas the erosive effect is mostly vertical. On the other hand, in arid zones evaporation is high, precipitation is low, and weathering is extremely slow due to the lack of water. In artic regions weathering is also very slow because the low temperature that prevails retards the chemical reaction rates.

The effect of climate on geomorphologic processes is felt as we move from one climatic zone to another. Variations in the extent of rainfall from one point to another within a given climatic zone can also produce drastically different landscapes. Similarly, variations in evaporation, infiltration, and runoff within a single climatic zone contribute to create differences in landscape.

The erosive work of streams is greatly affected by climatic changes. A similar statement can be made in regard to the structuring of the shoreline by wave action, although perhaps this last case is less obvious.

Some of the most important climatic factors that affect geomorphologic processes are:

(a) extent and distribution of precipitation,
(b) extent and distribution of evaporation,
(c) wind velocity patterns,
(d) temperature variations,
(e) storm patterns,
(f) depth of frost penetration, and
(g) humidity levels.

The effect of each of these factors on various geomorphologic processes is difficult to examine. For instance, an increase of temperature seems to accelerate the process of chemical weathering since it tends to increase the speed of such reactions. However, at the same time such an increase in temperature will produce an increase in evaporation and therefore reduce the amount of water available for weathering.

Numerous studies have attempted to show the relationship between climate and various geomorphologic processes.[26–28] Much work has been done on the effect of climate on weathering, especially in the humid tropics.

2.6 SOIL EROSION

One of the most important exogenetic processes that re-shapes land forms is erosion. This term comes from the latin "erodere" which means to reduce something in size or gnaw it away. Erosion of the soil is a very pressing problem to farmers, as well as agricultural and construction engineers. In addition, soil erosion plays an important role in the development of a drainage basin.

Soil erosion involves the action of water (surface channels, ocean waves, groundwater, and glaciers) as well as that of the wind. Weathering while not a part of the erosion process tends to make it proceed at an easier pace.

River erosion has already been discussed in a previous section (2.3). Most soil erosion, however, is not the result of the work of rivers or the wind, but involves the removal of loose soil material by percolating rainwater, downslope transport of the removed material by sheet flow, and deposition of the eroded material further downslope.

Soil erosion is considerably reduced by the presence of a vegetal cover. A good grass sod will reduce the impact force of the particles of rain impinging on a given parcel of soil, thus reducing the soil-tearing ability of the rain. Along the side of a hill the spots most affected by erosion are the steeper points about mid-length downslope[5] (zone ab in figure 2.1).

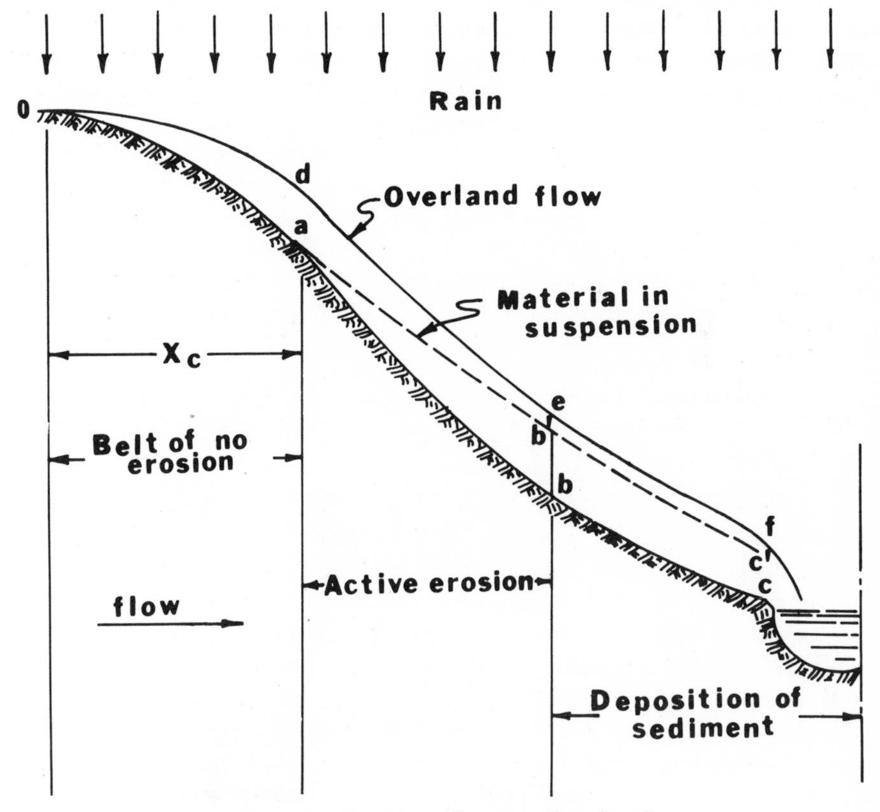

Figure 2.1. Partial profile of a valley slope.[5]

Horton[5] has quantitatively considered the process of soil erosion on a typical stream valley slope as shown on figure 2.1. Here oabc represents the soil surface profile and ab is its steeper portion. The surface of sheet flow during a typical thunderstorm is depicted by odef. The half-profile of the valley slope consists of three regions, namely, the belt of no erosion (oa), the active erosion zone (ab), and the region of deposition of sediment (bc).

In the belt of no erosion the sheet flow energy is insufficient to tear up the soil and therefore no erosion takes place. As the sheet flow approaches point "a" erosion begins to take place and proceeds down to point "b" continually removing material. After the flow sheet reaches point "b" no more material can be picked up unless some soil is removed from the water and re-deposited. Whatever soil remains in solution is carried past "c" into the stream and may be transported many miles downstream by the force of the river's current.

Cloudburst floods are an important factor in soil erosion. Such floods can carry tremendous amounts of solids in water suspension and lead to a high degree of erosion. While a good grass sod usually tends to reduce erosion of the soil, this is not necessarily the case during an intense storm or flood as illustrated by figure 2.2.[5] At the beginning of the rain the grass blades stand-up (figure 2.2a) and tend to slow down the sheet flow and thus increase the depth of surface retention. As this depth increases (b) the grass blades begin to flatten and erosion can start. Eventually, (c) a piece of sod is torn, and rolls down tearing up more soil downslope.

Once the soil is torn from a given parcel it will be transported downslope by the water. Such material can be carried in four different ways[29]

(a) in suspension,
(b) as dissolved solids,
(c) by saltation (grains hopping up and down), and
(d) as bed load.

A given sheet of water can carry a limited amount of solids and plant debris. At spots downslope some of the debris may suddenly pile up and distort both the runoff and soil erosion pattern.[29]

Eventually, the eroded material is deposited either at the bottom of the valley slope or further down-the-river. Rivers transport and deposit tremendous amounts of solids. Rivers the size of the Niger in Africa and the Mississippi in the United States carry many million cubic feet of solids every year. These solids are eventually dumped into the ocean where they tend to form large deltas.

Soil erosion is an important problem to farmers since uncontrolled erosion can destroy the usefulness of a parcel of land for agricultural pur-

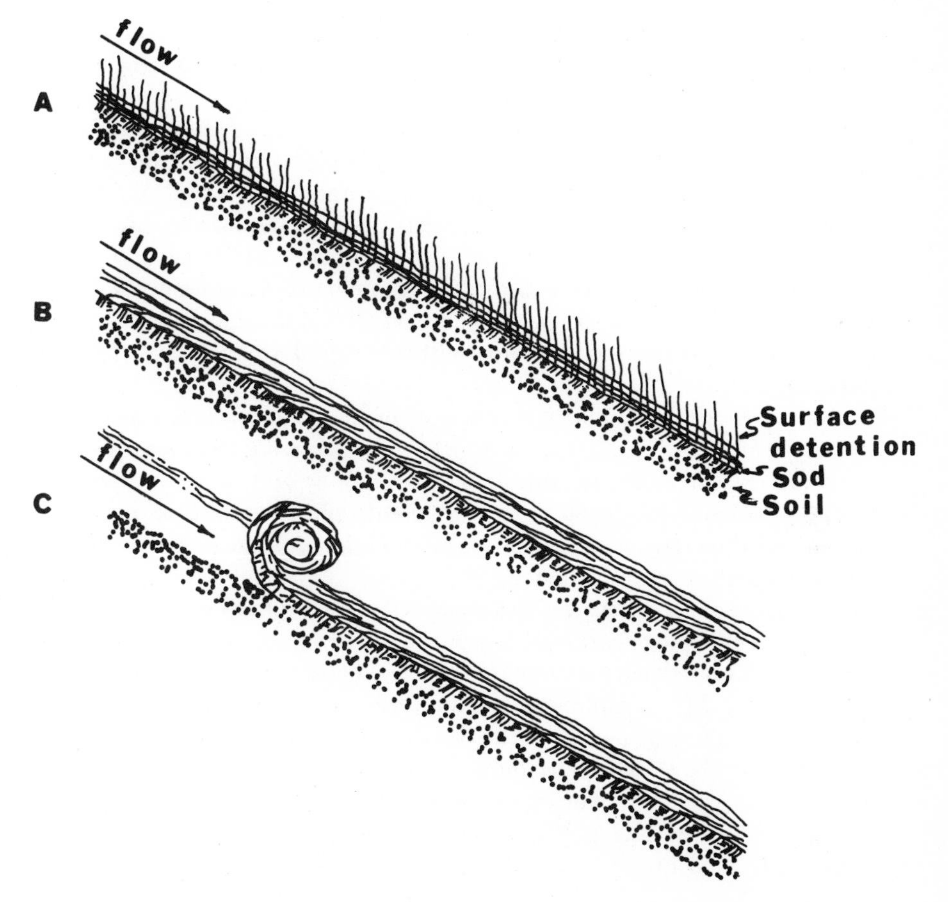

Figure 2.2. Erosion of sodded area initiated by the breaking down of grass cover in intense rains.[5]

poses. A good vegetal cover is one of the best protective measures that farmers can use to prevent unwanted erosion. Soil erosion can also affect highways and construction works. Dams, by affecting the transport phase of soil erosion, can in turn affect the soil erosion pattern in a drainage basin. Diversion and irrigation channels can also affect overall erosion patterns. Thus, the importance of proper design of all such works cannot be overemphasized.

2.7 MAN AS A GEOMORPHOLOGIC AGENT

In the previous sections we have examined the diversity of natural processes that in one way or another help to constantly re-shape the earth's surface. The effect of many such natural processes is greatly accelerated by the work of man.

From a geomorphologic standpoint man should be classified as both a builder and a destroyer. As examples of man's ability to build we can cite such feats as high-rise buildings, highways, rail lines, artificial islands, artificial bodies of water, bridges, tunnels, wells, mines, dams, etc. However, at the same time many examples of man-made spoilage can be cited, such as, the accumulation of litter, the dumping of mine wastes without considering the effect of such an action on the local ecology, and the restructuring of beaches and shores without regard to the ocean current patterns.

Part of the geomorphological influence of man is of an indirect nature, by virtue of accelerating naturally-occurring processes. For instance, by dumping excessive organic nutrients (see Chapter 3) into small quiescent bodies of water, man tends to accelerate eutrophication of such bodies of water and thus they turn into swamps much faster than by the action of nature alone.

In some instances, man has not been well aware or has neglected to consider the engineering characteristics of the soil in a given area. Such neglect can lead to very dangerous circumstances. The city of San Francisco is a case-in-point. Many buildings in San Francisco including some schools have been built right on top of the San Andreas Fault!

Aside from the danger of earthquakes in areas like San Francisco, serious problems can result in many areas from geomorphologic processes such as subsidence and slumping. Careful engineering planning can help avoid many such problems.

2.8 HAZARDS RESULTING FROM THE OCCURRENCE OF GEOMORPHOLOGIC PROCESSES

When geomorphologic processes occur rather suddenly they can cause a tremendous change in local environment and bring about a hazardous situation to all living things in the area. Numerous geomorphologic processes such as volcanic eruptions, intense storms, landslides, earthquakes, etc. can create chaos in a matter of hours and sometimes even minutes. Other geomorphologic processes such as subsidence, even though much slower than those above, can pose many problems especially to buildings and roads. To illustrate the nature of the hazards posed by some of these processes we will consider the effect of both subsidence and landslides. For a comprehensive description of other geomorphologic processes that lead to hazardous situations the reader should refer to the recent compilation on engineering geology problems by Lung and Proctor.[30]

Subsidence can be defined as a lowering of the land surface. Although subsidence has occurred all through geologic time as a result of natural causes, the work of man can specifically trigger or intensify subsidence in a local area. This was the case in Long Beach, California where subsidence occurred as a result of the withdrawal of oil from a reservoir under the city and in Mexico City where subsidence was triggered by excessive pumping of groundwater.[31]

The extent of subsidence resulting from the withdrawal of underground fluids[32,33] depends on the volume of fluid removed from within the compressible confining layers in the subsurface, the specific storage[4] of the underground strata, the rock compressibility, and the fluid's compressibility. Clay-rich or poorly-consolidated soils are prime spots for subsidence due to their usually high fluid content, and relatively uncompacted state.

Land subsidence resulting from the withdrawal of underground fluids may be halted by prohibiting such withdrawal or reducing the rate of extraction. In some areas injection of fluid into the subsurface may be needed in order to sufficiently reduce the rate of subsidence. This injection approach for the case of subsidence due to oil production consists basically in waterflooding the oil reservoir. Such an approach should not only help to stop subsidence but may act to increase oil production in the field being flooded.

Subsidence can also have hazardous indirect effects. For example, subsidence of coastal areas will increase their vulnerability to floods. Subsidence can also create cracks in the land surface. Such an occurrence is suspected to have been the cause for the break-up in 1963 of Baldwin Hill Reservoir in Los Angeles, California.[31]

Landslides can also lead to very hazardous situations. They can be very sudden or take place somewhat slowly. In either case they can be very destructive. A type of landslide that can affect construction of homes in many areas where shaly rocks occur is called successive rotational or slump slide. Sharpe[34] defines slumping as the "downward slipping of a mass of rock or unconsolidated material of any size, moving as a unit or as several subsidiary units, usually with backward rotation on a more or less horizontal axis parallel to the cliff or slope from which it descends."

Slumping is a very common occurrence throughout the world. To illustrate the process we will consider the Illinois River Bluff near La Salle and Peru, Illinois. Heavy rains during both the spring and summer of 1970 acted to trigger the slumping of portions of the Illinois River Bluff south of La Salle and Peru. Such geomorphological activity led to serious construction problems.[35] Figure 2.3 shows a locality in the Illinois River area

Figure 2.3. Terracettes formed by slumping along small tributary of the Illinois River.[35] (*courtesy Illinois State Geological Survey*)

Figure 2.4. New home constructed along the bluff. Recent landslide obstructs the road. Fracturing and slumping are visible above and to the left of the house.[35] (*courtesy Illinois State Geological Survey*).

Figure 2.5. A road in Illinois damaged by slump activity and now abandoned.[35] (*courtesy Illinois State Geological Survey*).

where slump blocks led to the development of terracettes on both sides of the stream. Sealing, cribbing, and draining have been used to stabilize the home on the right side of the figure. When cracks appeared next to the foundation, construction on the home on the left side of figure 2.3 was stopped.[35]

Figure 2.4 shows the bluff slope two miles east of the mouth of Allforks Creek (Illinois). Scarps* are visible, especially on the left side of the figure. Active slumping is visible above and below the house lot. The slide to the left of the house buried the road below.[35] The one lane section of road opened when part of the slide was cleared is offset 20 feet from the original road!

Slumping, as pointed out in the preceding paragraphs, can cause major damage to both roads and homes. Figure 2.5 shows a road in Illinois damaged by slump activity and now abandoned. Figure 2.6 shows a major slump that developed on the slope just above the house on the left side of the figure. The flowage of material onto the back of the house can eventually cause considerable damage unless costly measures are undertaken.

Weathered shales are more prone to sliding than unweathered ones because the former tend to take up more rain water than the latter. Areas of heavy rainfall and poor drainage are most favorable for slumping than areas of average or low rainfall and well-integrated drainage patterns.

Visibility of slump structures from the ground can be hindered by the presence of foliage. One characteristic feature that helps in recognizing unstable slopes is the presence of tilted trees (see figure 2.7). Such a feature is commonly called "a staggering forest". The tilted trees ride the slump block downslope and thus control in part its size since they serve to hold together large masses of sliding material.

Adequate recognition of potentially unstable land masses can be very helpful in avoiding hazardous situations created by carelessly planned construction of homes and roads. Expansion of homesite developments into potentially unstable slopes should be strongly discouraged by the government agencies regulating such developments in a given area.

2.9 REGIONAL OBSERVATION OF LAND FORMS

In previous sections of this chapter we have discussed how the action of various geomorphologic processes works to create land forms, man's

---

* Scarps can be defined as natural exposures of fresh earth and rock material uncovered as a result of slumping.

Figure 2.6. Large well developed landslide above buildings on the slope of a tributary valley.[35] (*courtesy Illinois State Geological Survey*).

Figure 2.7. Poor drainage is a conspicuous feature associated with slumping. Such activity causes cracks and depressions near the scarp faces and frequently disrupts normal flow patterns.[35] (*courtesy Illinois State Geological Survey*).

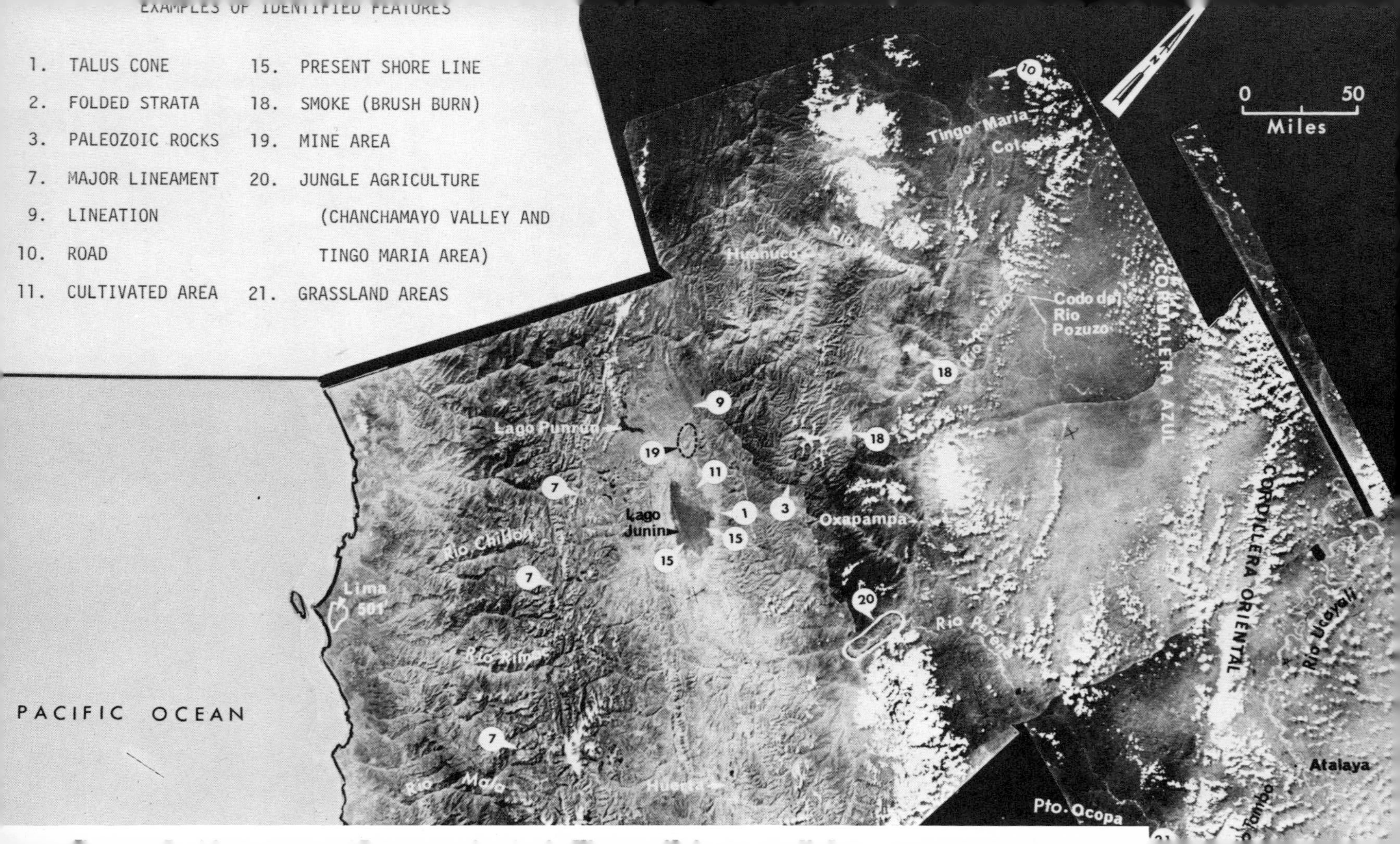
EXAMPLES OF IDENTIFIED FEATURES
1. TALUS CONE
2. FOLDED STRATA
3. PALEOZOIC ROCKS
7. MAJOR LINEAMENT
9. LINEATION
10. ROAD
11. CULTIVATED AREA
15. PRESENT SHORE LINE
18. SMOKE (BRUSH BURN)
19. MINE AREA
20. JUNGLE AGRICULTURE (CHANCHAMAYO VALLEY AND TINGO MARIA AREA)
21. GRASSLAND AREAS
0 50 Miles
Tingo Maria
Cole
Rio
Huanuco
Codo del Rio Pozuzo
Rio Pozuzo
CORDILLERA AZUL
CORDILLERA ORIENTAL
Rio Ucayali
Atalaya
Pto. Ocopa
Rio Perene
Oxapampa
Lago Punrun
Lago Junin
Rio Chillon
Rio Rimac
Rio Mala
Lima 501'
PACIFIC OCEAN

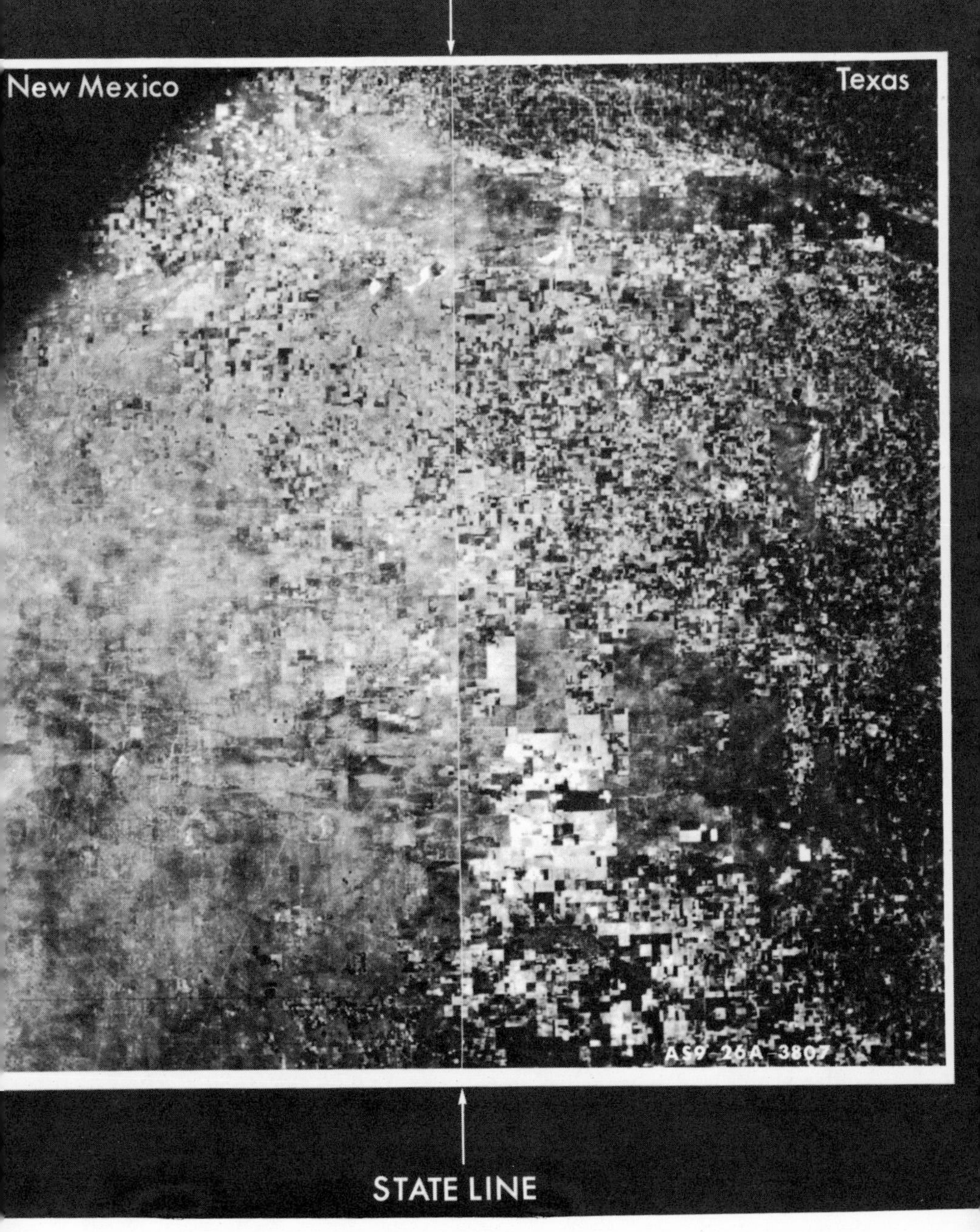

Figure 2.9. Photograph of Texas and New Mexico taken from Apollo 9 (*courtesy of U.S. Department of the Interior, Geological Survey*).

influence on land forms, and the sometimes hazardous interaction between geomorphologic processes and man. Improved understanding of all these areas has been achieved by the use of space photography and other remote sensing techniques. Space photographs offer an unsegmented view of a region under uniform lighting conditions and allow detection of prominent geologic features and even small changes in landscape.[36]

The National Aeronautics and Space Administration (NASA) and the United States Geological Survey (USGS) have taken a lead in the use of space photography to improve our understanding of land forms. For instance, figure 2.8 shows a photo-image map of Peru constructed by putting together twelve space photographs rectified to their true vertical position. Such a mosaic shows each feature in its true geographic position because each rectified photograph was essentially orthographic.

In addition to space photography other remote sensing techniques are very useful in the study of land forms. These include radar, sonar, spectrophotometry, infrared and ultraviolet sensors, etc.[37] Some of the objectives of remote sensing of land forms are:

1. to detect prominent geologic features;
2. to detect pollution problems and their relation to landscape;
3. to construct soil distribution maps;
4. to explore for mineral deposits;
5. to detect faults and fractures;
6. to construct tectonic maps;
7. to construct topographic maps;
8. to classify land forms;
9. to determine the relation between currents, tides, and sediment distribution; and
10. to identify hydrologic phenomena and evaluate moisture content of the soil.

Viewing land forms from space can also be helpful in policy making.[38] Figure 2.9 shows the marked contrast (visible from an altitude of 119 nautical miles) between the states of Texas and New Mexico reflecting the difference in land usage which in part results from differences in the laws regulating the extraction of groundwater in each of the two states.[38] This picture was taken from Apollo 9 on March 12, 1969.

With recent advances in space technology it is not far fetched to think that remote sensing techniques could be used in the future to forecast geomorphologic processes that may be hazardous to man and/ or other living beings.

2.10 *REFERENCES*

1. Thornbury, W. D. (1969) "Principles of Geomorphology", John Wiley and Sons, New York, N.Y., second edition.
2. Tarr, R. S. and Martin, L. (1914) "College Physiography", The Macmillan Company, New York, N.Y.
3. Playfair, J. (1802) "Illustrations of the Huttonian Theory of the Earth", William Creech, Edinburgh (American edition, 1856– University of Illinois Press).
4. Deju, R. A. (1971) "Regional Hydrology Fundamentals", Gordon and Breach Science Publishers Inc., New York, N.Y.
5. Horton, R. E. (1945) *Erosional Development of Streams and Their Drainage Basins*, Geological Society of America Bulletin, volume 56, pp. 275–370.
6. Frye, John C. (1971) "A Geologist Views the Environment", Environmental Geology Notes #42, Illinois State Geological Survey, Urbana, Illinois.
7. Anonymous (941–982) "The Discourses of the Brothers of Purity", Ikhwan es Safa, in arabic, four volumes.
8. Said, Rushdi (1950) *Geology in Tenth Century Arabic Literature*, American Journal of Science, volume 248, pp. 63–66.
9. Gilluly, J., Waters, A. C., and Woodford, A. O. (1959) "Principles of Geology", W. H. Freeman and Company, San Francisco, Cal., second edition.
10. Hutton, James (1788) *Theory of the Earth; or an Investigation of the Laws Observable in the Composition, Dissolution, and Restoration of Land Upon the Globe*, Transactions of the Royal Society of Edinburgh, volume 1.
11. Davis, W. M. (1889) *The Rivers and Valleys of Pennsylvania*, National Geographic Magazine, volume 1, pp. 183–253.
12. ——— (1899) *The Geographic Cycle*, Geographical Journal, volume 14, pp. 481–504.
13. ——— (1899) *The Peneplain*, American Geol., volume 23, pp. 207–239.
14. ——— (1954) "Geographical Essays", Dover Publications, Inc., New York, N.Y. (reprinted from the 1909 original).
15. Penck, W. (1924) "Die Morphologische Analyse. Ein Kapitel der Physikalischen Geologie", Stuttgart, Geog. Abh. 2 Reihe, heft 2.
16. ——— (1953) "Morphological Analysis of Landforms", St. Martin's Press, New York, N.Y. (a translation).
17. Hack, J. T. (1960) *Interpretation of Erosional Topography in Humid* Temperate Regions, American Journal of Science, (Bradley) volume 258-A, pp. 80–97.

18. Chamberlin, T. C. and Salisbury, R. D. (1904) "Geology", Henry Holt and Co., New York, N.Y.
19. Leopold, L. B., and Wolman, M. G. (1957) "River Channel Patterns: Braided, Meandering and Straight", U.S.G.S.-Prof. Paper 282-B.
20. Sandborg, A. (1956) *The River Klarälven, a Study of Fluvial Processes,* Geografiska Annaler, volume 38, pp. 127–316.
21. Deju, R. A. and Bhappu, R. B. (1966) "A Chemical Interpretation of Surface Phenomena in Silicate Minerals", N. M. Bur Mines, Min. Resour., Circular 89.
22. ——— (1971) *A Model of Chemical Weathering of Silicate Minerals,* Geological Society of America Bulletin, volume 82, pp. 1055–1062.
23. Jenny, H. (1941) "Factors of Soil Formation", McGraw Hill Book Co., Inc., New York, N.Y.
24. Tyrrell, G. W. (1931) "Volcanoes", T. Butterworth, London, England.
25. Ashley, G. H. (1931) *Our Youthful Scenery,* Geological Society of America Bulletin, volume 42, pp. 537–546.
26. Loughnan, F. C. (1969) "Chemical Weathering of the Silicate Minerals", American Elsevier Publishing Company, Inc., New York, N.Y.
27. Crowther, E. M. (1930) *The Relationship of Climate and Geological Factors to the Composition of the Clay and the Distribution of Soil Types,* Proc. of the Royal Soc., volume 107, pp. 10–30.
28. Krynine, P. D. (1936) *Geomorphology and Sedimentation in the Humid Tropics,* American Journal of Science, volume 232, pp. 297–306.
29. Gilbert, G. K. (1914) "The Transportation of Debris by Running Water", U.S.G.S.—Prof. Paper 86.
30. Lung, R. and Proctor, R. (eds.) (1966) "Engineering Geology in Southern California", special publication of the Los Angeles Section of the Assoc. of Engineering Geologists, Arcadia, Cal.
31. Marsden, Jr., S. S. and Davis, S. N. (1967) *Geological Subsidence,* Scientific American, volume 216, pp. 93–100.
32. Poland, J. F. (1961) "The Coefficient of Storage in a Region of Major Subsidence Caused by Compaction of an Aquifer System", U.S.G.S.-Prof. Paper 424-B.
33. Domenico, P. A., and Mifflin, M. D. (1965) *Water From Low Permeability Sediments and Land Subsidence,* Water Resources Research, volume 1, pp. 563–576.
34. Sharpe, C. F. S. (1938) "Landslides and Related Phenomena, A Study of Mass-Movements of Soil and Rock", Columbia University Press, New York, N.Y.
35. DuMontelle, P. B., Hester, N. C. and Cole, R. E. (1971) "Landslides Along the Illinois River Valley South and West of La Salle and Peru,

Illinois", Environmental Geology Notes #48, Illinois State Geological Survey, Urbana, Illinois.

36. Carter, W. D. (1971) *ERTS—A A New Apogee for Mineral Finding*, Mining Engineering, volume 23, pp. 51–53.
37. Malin, Jr., H. M. (1971) *Pollution Detection by Remote Sensing*, Environmental Science and Technology, volume 5, pp. 676–677.
38. Fischer, W. A. (1971) *Uses of Space Observations of the Earth-The EROS Program of the Department of the Interior*, II Interamerican Conf. on Materials Technology Proceedings, Mexico, D.F. pp. 440–444.

Chapter 3

WATER RESOURCES

3.1 *INTRODUCTION*

As we pointed out in the first chapter, one of the most valuable resources that we have is water. Unfortunately, for many decades people considered it as simply "a natural thing to have," and "of little economic value." It was not until very recently that people began to be concerned with water pollution problems and mismanagement of water resources.

In order to integrate the efforts of the United States Federal Government in the area of pollution abatement, the Environmental Protection Agency (EPA) was created in 1970. Some of the most important priorities of this agency are the problems of water resources. The Federal Government presently supports research programs in many areas related to water, such as water quality research, studies on urban water problems, development of desalting techniques, etc. The Federal Government also administers grants to small communities for the construction of waste treatment works. Expenditures by the Federal Government in the area of water pollution have risen from forty four million dollars in 1961 to approximately four hundred and sixty five million dollars in 1971.

Given the great importance of water, an entire chapter will now be devoted to continental water while the following chapter treats the oceans. Our discussion here will cover the sources of water and the conditions under which surface and subsurface waters are found, the methods for regionally evaluating water resources, and the prospects of supplying water in future years. In the area of water pollution we will discuss water quality criteria, some of the most important and dramatic water pollution problems, and the techniques for wastewater treatment and re-use. In addition, we will discuss the considerations that should be given concerning the availability and quality of water prior to establishing an industry in a particular locale. Finally, we will take a look at the future and attempt to visualize the water problems that must be tackled in the coming years.

An entire book could have been written on the subject of water, however, this is not our intent. For a more detailed treatment of water the reader can consult any of several books available on this specific subject.[1–4]

Figure 3.1a. Sailboat on beautiful Grand Lake Colorado (*courtesy The Denver Post*).

Figure 3.1b. O'Traver Lake near Salida, Colorado (*courtesy The Denver Post*).

Figure 3.1c. A superbly beautiful Beaver dam on the upper north fork of the Platte River (*courtesy The Denver Post*).

Figure 3.1d. The beauty of northern Saskatchewan is augmented by the numerous bodies of water in the area.

3.2 *SOURCES OF WATER*

As we pointed out in the previous section water is one of the most important natural resources available to mankind. Man can not survive very long without water and neither can plants nor animals. Therefore, one should try to preserve and use, as efficiently as possible, this valuable resource.

Civilizations from times inmemorial have been concerned with the discovery and utilization of sources of water. Nearly 4000 years ago the Chinese were actively drilling water wells. Even the wise King Solomon of Biblical fame dedicated a considerable amount of his time to the design of aqueducts and the development of the water resources of his kingdom.

Basically, there is a fixed amount of water in nature, and this water can be found in different points in space, (see hydrologic cycle in the following section) at different times. As far as man is concerned there are four possible sources of water, which are:

1. the oceans,
2. continental bodies of surface waters,
3. groundwater, and
4. recycled wastewater.

The oceans constitute a huge reservoir of water occupying 70.8% of the earth's surface to an estimated average depth of 12,500 feet.[5] Even though ocean water comprises over 98% of the hydrosphere, the direct utilization of this water is limited due to its salinity. Nonetheless, some desalination plants have been installed in coastal cities in order to make ocean water suitable for consumption.[6] Many of these installations not only produce good quality water, but also salts of sodium, potassium, and magnesium, hydroelectric power, and bromine.[7–9] Thus, many byproducts can result from desalination operations in addition to the water.

Surface waters are also a valuable source not only from the standpoint of consumption, but also as an asset for recreational activities. Rivers and lakes also enhance the natural panorama of a region (see figure 3.1).

Groundwater is also a very important source of good quality water. In certain areas where there is little runoff many municipalities and industries depend greatly, if not entirely, on groundwater.

Finally, a source of water that is acquiring more importance each day is reclaimed wastewater. For many applications where a lower quality requirement is acceptable, treated wastewater can be used or a recycling process can be carried out.

When the sources of water in a given area are to be evaluated the hydrologist must first carefully establish the water needs of the area as far as quantity and quality are concerned. Then, he can proceed to use his hydrologic knowhow to attempt to determine which sources to tap.

3.3 *THE HYDROLOGIC CYCLE*

Water in nature is constantly on the move. Even early scientists like Plato in Greece and Marcus Vitruvius in Rome realized this fact and in their own crude ways constructed a picture of the hydrologic cycle.

Most of the early theories about the hydrologic cycle depicted water flowing through mysterious subterranean channels connecting the bottom of rivers and lakes to the oceans. These channels can be visualized as the earliest concept of aquifers. Also, they believed that the total amount of rainwater on the earth was not enough to balance the water discharged by rivers and springs, which we now know is not true.

While these early theories were rather rudimentary they were actually not so far off from our present idea concerning the hydrologic cycle.

The modern hydrologic cycle (see figure 3.2) includes the processes of precipitation, evaporation, sublimation, transpiration, runoff, infiltration, and groundwater flow. These processes occur in three regions:

(a) the region composed of the atmosphere and the surface of the earth,

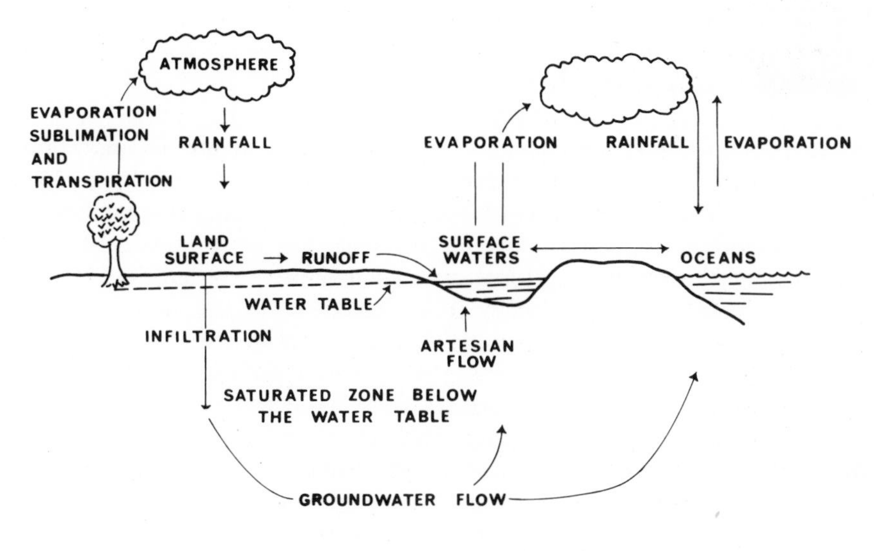

Figure 3.2. Hydrologic Cycle

(b) the region of unsaturated groundwater flow, and
(c) the saturated flow region.

These last two are separated by the *water table.*

The basis of water supply is to find which of the above regions to tap as a source of water in a particular situation.

Aside from its didactic importance the hydrologic cycle furnishes us with a tool to determine in a given area the extent of a given hydrologic process knowing the extent of all other hydrologic processes taking place there. Such calculations are called *hydrologic inventories.* There are three types of inventories:

(a) Global Inventory:

Total Worldwide Precipitation = Total Worldwide Evaporation, Transpiration and Sublimation + Change in Subsurface Storage

(b) Continental Inventory:

Precipitation = Evaporative Processes + Change in Subsurface Storage + Runoff

(c) Basinwide Inventory:

Basin Inflow + Precipitation = Evaporative Processes + Change in Subsurface Storage + Runoff Outflowing the Basin

3.4 *MOVEMENT OF SURFACE WATERS*

As pointed out in section 3.2 surface waters constitute an important portion of the world's water resources. Water flowing on the earth's surface is not only an important source of water supply but is also a very important geomorphological agent. Rivers, for instance, are most prominent geomorphological agents and possess tremendous erosive capacities.

In studying surface hydrology it is convenient to use as a frame of reference the concept of a *basin.* This concept will be used here much in the same way as a mathematician would use a cartesian or any other coordinate system. In simple terms a river basin can be defined as the total area where runoff in the direction of the river takes place. All points in the basin contribute runoff water to the river and its tributaries and undergo geomorphological re-shaping due to water movement in the basin.

Runoff and river discharge are probably the most obvious types of surface water movement. Other types of water movement occur between the land surface and the atmosphere and are not important from an erosive standpoint but greatly affect the quantity of water in the basin. These are precipitation, transpiration, sublimation, and evaporation.

The effect that movement of surface waters has in re-shaping the earth's surface is tremendous. Rivers are very effective erosion agents. Their erosive capacities are a function of the volume of water they discharge and their suspended load. The erosive capacity of a river can be related to both the maximum size of particle of a given type that it can carry under the existing flow regime, and the total particle load it can carry.

The load carried by a river is mostly the result of rainwash and downslope movements over the basin. This load is enhanced by particle abrasion from bed and banks. As some particles roll along the river bed they can strongly abrade the bedrock. The suspended and dissolved load of particles can also affect a river's erosive capacity.

River discharge is an important concept from both a water supply and a geomorphological standpoint. It can be defined as the total amcunt of water carried by the river past a given point. Mathematically, it is given by the formula

$$Q = \bar{v}A \tag{3.4–1}$$

where Q is the total discharge, $\bar{v}$ is the average river velocity, and A is the cross-sectional area at the point of measurement. If the velocity is a variable, then,

$$Q = \int v \, dA \tag{3.4–2}$$

Movement in surface water bodies such as rivers occurs as a result of gravity. The velocity of such flow, v, is related to many hydrologic factors. The square of the velocity is directly proportional to the river's cross sectional area and its slope, and inversely proportional to the wetted perimeter and the stream's roughness coefficient.

Movement of water is important in keeping the water fresh. This can be shown for the case of many lakes where among other things due to the lack of movement the lake water tends to develop plankton and serves as breeding ground for insects and bacteria.

Movement of surface waters is seasonally affected by climatic conditions. When a surface-hydrology study is carried out in order to establish water supply sources one must begin by carefully determining the total stream flow available under varying climatic conditions. The purpose of such a task is to first show if storage of surface waters is necessary in the basin under study, and second, if storage is needed one must determine the extent to which it is essential.

Figure 3.3a. Rivers meander around in beautiful contorsions such as shown here (*courtesy of The Denver Post*).

Figure 3.3b. This figure shows the upper middle fork of the Platte River as it passes by a possible source of pollution, namely, some of Colorado's old mines (*courtesy of The Denver Post*).

Figure 3.3c. This figure shows a river polluted with detergent foam. This picture was taken near Glenwood Canyon (*courtesy The Denver Post*).

Figure 3.3d. Suburban effluent pollutes the Platte River near Denver (*courtesy The Denver Post*).

It should be by now clear to the reader that water in nature is always on the move, and that this movement may greatly vary. It is sometimes fast while other times it can be very slow. Precipitation, evaporation, transpiration, and sublimation are quite potent in some areas while they may be negligible in others. The same is true of runoff and river flow, as discussed in the preceding paragraphs.

Precipitation represents water gain in a basin while evaporation, transpiration, and sublimation represent water movement out of the basin, that is, water losses to the atmosphere. All these processes are very much interrelated and data about them are needed in order to evaluate surface water supply sources in a basin. Runoff and river discharge data are also important. Actually, what all these measurements represent is not only the amount of water being moved, but also the speed of motion, and its extent. Precipitation, evaporative processes, and runoff can be measured in various ways (see following section). These parameters are commonly expressed as a fraction or a few inches per day. River discharge, on the other hand, is expressed in terms of cubic feet/sec or gallons per minute, while stream velocity is given usually in terms of miles per hour. Most rivers have velocities between three and five miles per hour, although, these values may be exceeded during flood stages.

There has always been something poetic about the movement of a river that has greatly attracted man's attention. Maybe it is the force of its flow or the softness and gentleness of bouncing waters, but there are few natural wonders that can compare in beauty and grandeur to moving water (see figure 3.3) and it is a shame that such bodies of water have sometimes become polluted. Figure 3.3 also shows the detrimental effects of river pollution.

3.5 *SURFACE WATER MONITORING*

Water is always on the move. We saw in section 3.3 how water moved from one point to another in the hydrologic cycle. We saw in the previous section that the movement of surface waters causes tremendous geomorphological changes and practically re-shapes the earth's landscape. In addition, it should be noted that the amount of water moving between any two points in the hydrologic cycle is a most important quantity, that needs to be known, in order to make a hydrologic inventory.

There are four types of processes that one needs to know and thus monitor in order to have adequate hydrologic inventories in a given basin. These are:

1. precipitation,

2. water losses to the atmosphere
   a. evaporation,
   b. transpiration, and
   c. sublimation,
3. runoff and river discharge, and
4. infiltration.

For a complete discussion of all these processes the reader can refer to an advanced hydrology book.[1,10–12] Here it will suffice to give a short description of each of the above processes and the way they can be measured. River discharge has already been treated in the preceding section.

Precipitation

Precipitation is measured using rain gages such as depicted in figure 3.4 which shows a rain gage and recorder. Basically, there are two types of gages, a recording and a non-recording one. The standard non-recording gage consists of a circular cylinder eight inches in diameter that acts as an overflow reservoir, a funnel type receiver also eight inches in diameter, and a measuring tube whose cross sectional area is $\frac{1}{10}$th that of the receiving tank. Rainfall is measured in the graduated measuring tube to the nearest $\frac{1}{100}$th of an inch.

Sometimes it is necessary in a regional investigation to calculate the average precipitation per year for a given basin. To do this, one needs data from various stations in the basin. Then, the average precipitation can be computed using the Thiessen or Isohyetal methods.[1,10]

Evaporation and Sublimation

These two quantities can be measured in the field using evaporating pans. For instance, to measure evaporation near a lake one stations in its vicinity a Weather Bureau Class A Evaporating Pan. This pan is made out of galvanized, unpainted iron, and measures two feet in radius and ten inches in depth. It usually sits on a wooden frame one foot above ground level. By establishing a pan coefficient, which is the ratio of lake to pan evaporation in a given time, one can calculate lake evaporation by simply measuring evaporation in the pan and knowing the equivalence between lake and pan.

Transpiration

Transpiration is usually measured in the laboratory by determining the change in weight per time for an enclosed plant.

Figure 3.4. This is a recording rain gage useful for long-term, continuous recording of heavy rainfall. Rain is funneled from the collector into a storage tank. A float resting on the collected water operates the recording mechanism to provide a permanent strip chart record of accumulated rainfall versus time. An oil layer may be placed on the water to reduce evaporation and increase accuracy (*courtesy Leupold & Stevens Inc.*).

Runoff

Runoff is that portion of the precipitation that does not filter through the soil or is lost by evapotranspiration, but runs through the ground to replenish bodies of surface water. Its analysis requires the use of streamflow records. These records are kept in the United States by the United States Geological Survey and the various state offices. In other countries equivalent agencies keep similar records. Streamflow records are obtained by determining at a given station the elevation of the water level and constructing rating curves.[1]

Infiltration

As we previously pointed out, a certain portion of the water that precipitates in a basin filters through the soil to replenish underground reservoirs. This infiltration can be estimated from the type of soil. Infiltration data is usually expressed in terms of infiltration capacity which is the maximum rate at which a given soil can, under given circumstances, sorb precipitation.

The monitoring of the above parameters is a very important task and should not be taken lightly. Such measurements furnish the hydrologist with one of the most valuable tools that he has in order to evaluate the water resources in a given area.

3.6 *AQUIFER GEOLOGY*

In addition to surface waters which have been discussed in the previous sections one can in many instances utilize water stored in the subsurface. As will be shown later in this chapter groundwater can play a most important part in the economy of a region.

Water is distributed in the subsurface in two main regions called the region of aeration and the saturated region (see figure 3.5). These two regions are separated by the water table. Above the water table is the region of aeration where some pores are not filled with water but contain air. The saturated region is below the water table and, in it, all the pores are filled with water. This is the most important region from a water supply standpoint. This region is actually a huge reservoir of water under hydrostatic pressure.

The region of aeration is one of suspended water. It is usually considered to be divided into three zones. The uppermost zone is the soil and root belt. This belt contains mostly soil moisture. The extent of the water held in this belt will be closely dependent on the type of soil and vegetation.

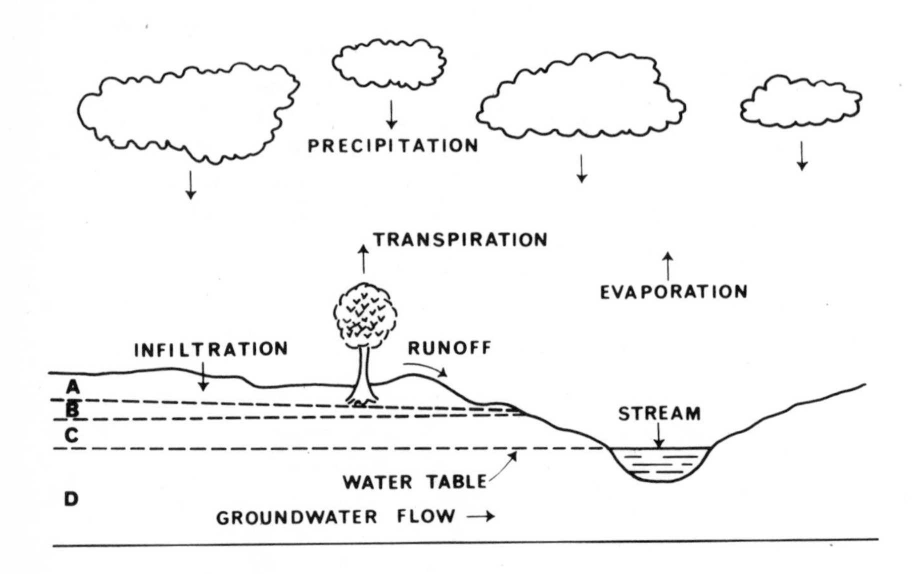

ZONES OF WATER IN SUBSURFACE:

REGION OF AERATION
- A. SUBSOIL
- B. VADOSE ZONE
- C. CAPILLARY ZONE

SATURATED REGION
- D. SATURATED GROUNDWATER FLOW

Figure 3.5. Vertical distribution of subsurface waters.

Immediately below the soil and root belt is the vadose zone. In this zone water is held in suspension by molecular attraction and capillarity. Water in the vadose zone is not easily recoverable. The lowermost zone of the aerated region is a capillary fringe located directly above the water table. In this zone water is suspended as a result of capillarity acting against gravity.

It is clear that from a water supply standpoint the most important region is that below the water table. Water filters into the saturated region and fills aquifers. These are formations with sufficient porosity to store water and sufficient permeability to allow its movement. Most aquifers are supplied by fresh-water recharge except those containing connate water. Connate water is the salt water which was trapped in the pores of a sediment as this sediment was uplifted by a geologic process.

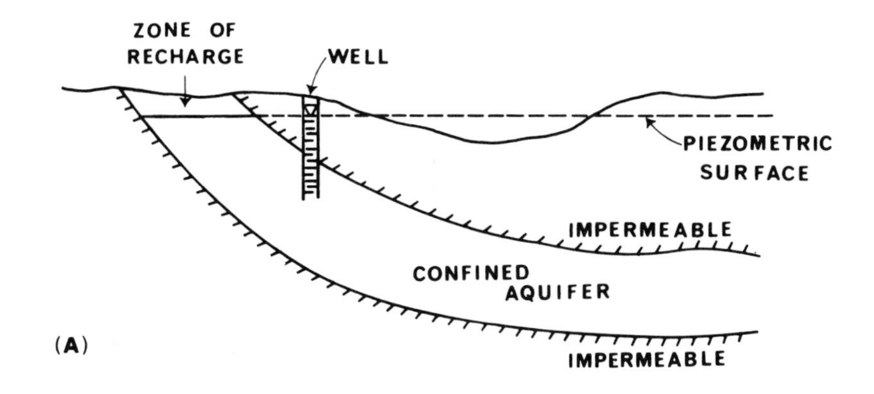

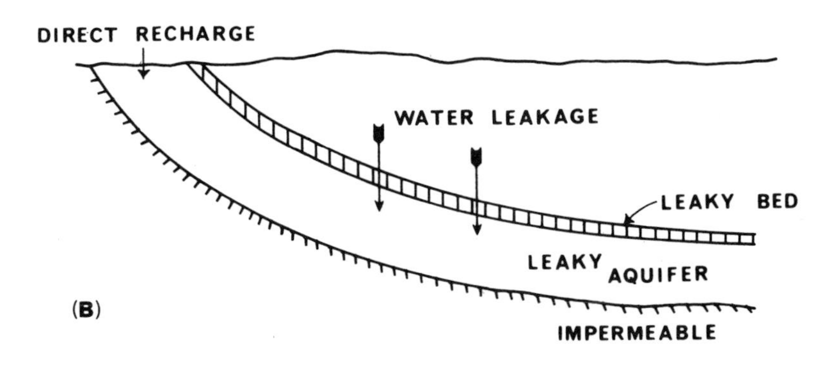

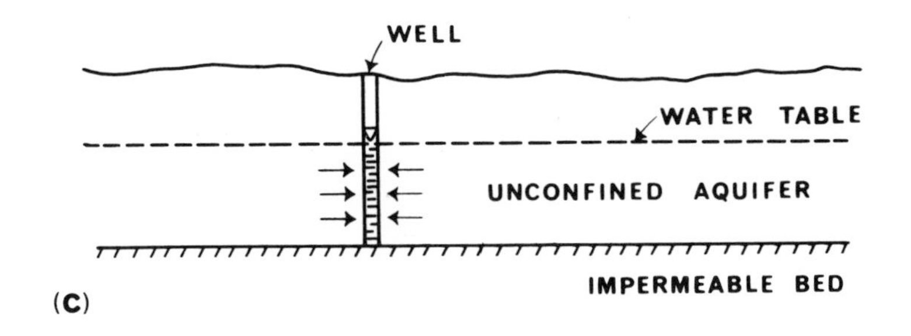

Figure 3.6. Types of aquifer. (A) confined; (B) leaky; (C) unconfined.

There are many formations which can act as aquifers. Most important of these are unconsolidated formations of quaternary age such as alluvial sands and gravels and deltaic deposits. In addition, some important aquifers are in older, consolidated formations such as permeable sands, fractured rocks of volcanic origin, and limestone formations.

From a hydrologic standpoint aquiferous formations are divided into three types (see figure 3.6) and, in each of these, groundwater movement must be described in a different fashion. The first type of aquifer is the confined case depicted in figure 3.6A. Such aquifers are separated above and below by impermeable beds. Water enters them only at the recharge area. A second type is the leaky aquifer (see figure 3.6B). Such an aquifer is usually overlain by a semipermeable or leaky bed which permits some vertical infiltration of water. The final type of aquifer is the unconfined or free case which is depicted in figure 3.6C. In this case the upper boundary of the aquifer is the water table.

It is not uncommon to have in a given region a combination of the above types. For example, there could be an unconfined aquifer in the overburden and a confined one further below.

3.7 *STORAGE AND MOVEMENT OF GROUNDWATER*

In the previous section we described the types of aquiferous formations that allow both movement and storage of groundwater. Such movement of water through a porous medium has been the subject of study since the XIX Century. Darcy (1856)[13] published the results of research that led to the law of flow through a porous medium that today bears his name. His law states that the total discharge, Q, through an aquifer is given by the formula

$$Q = -KA \frac{\mathrm{d}h}{\mathrm{d}l} \tag{3.7–1}$$

where K is a constant termed hydraulic conductivity, A is the aquifer's cross sectional area and $\mathrm{d}h/\mathrm{d}l$ is the hydraulic gradient.

The hydraulic conductivity (sometimes termed coefficient of permeability) depends on both the nature of the fluid and the type of porous medium. Some authors prefer to separate the dependency on the fluid from that on the formation. This can be accomplished since

$$K = \frac{k\rho g}{\mu} = kC \tag{3.7–2}$$

where k is the intrinsic permeability, C is a term that depends only on the fluid, ρ is the fluid density, μ its viscosity, and g the acceleration of gravity. The intrinsic permeability depends only on the porous medium and is defined as the ease of flow through a given formation. The intrinsic permeability is directly proportional to the square of the average pore diameter and it is influenced, as shown by Krumbein and Monk[14] by variations in mineral composition, sphericity, angularity, grain texture, and crystallographic orientation.

The intrinsic permeability of a sediment can be measured both in the field and in the laboratory.[1] Laboratory measurements, however, have the drawback that they may not be fully representative of the formation under study. Field measurements can be done either by pumping tests or using tracers and are much more reliable.

The hydraulic conductivity has units of velocity while the intrinsic permeability is expressed as length square. The most common unit for expressing intrinsic permeability is the darcy which is defined by the equation

$$1 \text{ darcy} = \frac{(1 \text{ centipoise})(1 \text{ cm}^3/\text{sec})/(1 \text{ cm}^2)}{(1 \text{ atmosphere/cm})} \tag{3.7–3}$$

Approximately speaking 1 darcy equals 10^{-8} cm^2.

Another important parameter affecting groundwater movement is the formation porosity. Porosity can be defined as the fraction of the total volume of the formation that is made up of pores. There are formations where the porosity is very homogeneous, such as in sandy aquifers, and others where pores are few and very irregularly distributed, for example, volcanic rocks.

Porosity is important from a groundwater standpoint since it gives us an idea of the amount of water stored in a given aquifer or petroleum reservoir. The maximum storage in an aquifer is equal to full occupancy of all pores by the fluid, however, in many instances a number of pores are isolated and any fluid contained in them will not be drawn out by pumping. Thus, one is (as far as water supply is concerned) only interested in the fluid contained in interconnected pores. This gives rise to two new concepts, namely, specific yield and specific retention. The former gives us the fraction of the porosity that yields water (or oil) when the head is lowered, while the latter represents the isolated pores. Of course, the porosity will equal the specific yield plus the specific retention.

Porosity is greatly affected by geologic processes to which sediments are commonly subjected to during and after deposition. There have been numerous studies concerning the effect that geological processes such as compaction, dissolution, and cementation may have on porosity.[15]

In general, heterogeneous sediments have smaller porosities than homogeneous ones since in the former the fines tend to fill up the pores of the coarser material. The minimum porosity of such a mixture will occur when the fines completely fill all the pores of the coarser material.

A third parameter that must be considered in a groundwater study is the transmissivity of the aquifer (sometimes called transmissibility). Basically, the transmissivity[16] is equal to the hydraulic conductivity of a formation times its thickness. It expresses the ability of a water-bearing formation to conduct water. The transmissivity of an aquifer is commonly expressed in gallons per day per foot. Its dimensions are $[L^2/T]$. Pumping tests are the most efficient way of measuring transmissivities.

A fourth parameter useful in groundwater studies is the coefficient of storage introduced by Theis[17] in 1935. This coefficient usually designated by the letter S defines the volume of water that a unit decline in head releases from storage in a vertical column of the aquiferous formation of unit cross sectional area. The coefficient of storage includes both the fraction of storage attributable to expansibility of the water and that attributable to compressibility of the aquifer. Mathematically speaking the storage coefficient S is given by the formula[16]

$$S = \rho g b(\alpha + f\beta) \tag{3.7–4}$$

where ρ is the density of the fluid, g the acceleration of gravity, α the vertical compressibility of the granular skeleton, f the porosity, β the compressibility of the fluid, and b the average thickness of the aquifer.

The coefficient S is non-dimensional. To illustrate its meaning consider a sand where $S = 3.65 \times 10^{-1}$. This value of S means that 365 cubic feet of water will be released from storage under an area of 1000 square feet as the head decreases one foot.

Finally, there are two parameters that are necessary in order to describe storage in leaky aquifers. The first one is the leakage factor, B, which can be expressed mathematically by the equation

$$B = \left(\frac{T}{(K'/b')}\right)^{1/2} \tag{3.7–5}$$

where T is the aquifer's transmissivity, and K' and b' are respectively the hydraulic conductivity and the thickness of the leaky bed.

The second parameter needed to describe leaky aquifers is the leakage coefficient or leakance. This parameter is equal to K'/b'.

To give the reader an idea of the range of values of the above parameters, Table 3.1 shows typical values of porosity, permeability, transmissivity, storage, leakage factor, and leakance. These quantities are commonly termed formation constants.

TABLE 3.1.

Some typical values of porosity, intrinsic permeability, transmissivity, storage, leakage factor, and leakance.

Formation	*f (porosity) %*	*k intrinsic permeability (darcies)*
freshly deposited alluvial material	85	100–1000
recent river deposits	45–50	100–1000
quicksand	50	100
well-sorted sands	25–45	10–1000
gravel	20–30	10000
sand and silt	10–15	0.01–0.001
clean limestone	15–20	0.1–0.5
clayey limestone	< 5	0.0001 or less
clayey sands	< 20	0.001–1.0
clays	< 25	0.001 or less

Area or Type	LEAKY ARTESIAN AQUIFER*				SHALLOW BASIN*		
	T (gpd/ft)	S**	K′/b′ (gpd/ft³)	B (ft)	T (gpd/ft)	S	B (ft)
ROSWELL	1,400,000	0.00001	.001140	35,000	100,000	0.10	9,500
DEXTER	75,000	0.00005	.000061	35,000	100,000	0.10	40,000
ARTESIA	150,000	0.00005	.000240	25,000	100,000	0.10	20,000
LAKEWOOD	66,000	0.00010	.000097	26,000	100,000	0.10	36,000
RECHARGE AREA	75,000	0.05	—	—	—	—	—
MOST GOOD AQUIFERS	$10000 < T < 10^6$				$0.001 < S < 0.30$		

* Data from Hantush, M. S. (1957) *Preliminary Quantitative Study of the Roswell Ground-Water Reservoir, New Mexico*, NMIMT Publication, Socorro, N.M.

** Low storage values for this artesian aquifer are most probably due to the low compressibility that characterizes such limestone aquifers.

To now illustrate the importance of groundwater storage consider the following case. We have an unconfined aquifer of an average thickness of 50 feet and occupying an area of 25 square miles. If the specific yield of this aquifer is 0.20 then such an aquifer assuming zero recharge will be able to yield 1.05 billion gallons (US) of water for every foot that the head is lowered. This is a tremendous amount of water. Just as an illustration, such an aquifer would be able to supply 1.50 million gallons per day for almost 20 years with zero recharge and only a 10 feet lowering of the water table!

Movement of groundwater is as important a topic as is storage. In addition, the same parameters used above are needed in order to describe groundwater movement. For instance, the velocity of moving groundwater can be expressed in terms of the hydraulic conductivity and the hydraulic gradient by combining equations (3.4–1) and (3.7–1). Thus,

$$v = -K \frac{dh}{dl} \tag{3.7–6}$$

which is another form of Darcy's law.

In general, groundwater velocities are very small and the above law applies. In certain instances, as for example, for the case of water moving toward a fast pumping well, turbulence may be created and the above law does not fully apply.

This brings us to the subject of well flow. The most common way of utilizing groundwater is to extract it by means of a pump located in a well tapping the aquifer. The flow of groundwater into the well can be of two types: steady, and unsteady. By steady flow one means the case where the water head does not vary with time. Thus, for this type of flow $\partial h/\partial t = 0$. For unsteady flow the head varies not only in space but also in time.

Flow into a well can be described mathematically under a given set of circumstances, and for a given type of aquifer by an "equation of motion." Such an equation basically describes the variations in head and drawdown near the well with respect to both space and time, as a function of the aquifer's transmissivity and its storage coefficient. For example, the movement of groundwater into a well from a confined horizontal aquifer is expressed in three-dimensional cartesian coordinates as

$$\frac{\partial^2 h}{\partial x^2} + \frac{\partial^2 h}{\partial y^2} + \frac{\partial^2 h}{\partial z^2} = \frac{S}{T}\frac{\partial h}{\partial t} \tag{3.7–7}$$

The right-hand-side of this equation equals zero for the case of steady flow since the head does not change with time under steady conditions.

The above equation was solved for unsteady conditions by Theis[17] in 1935. He obtained an expression for the drawdown "s", (s = initial water level minus water level at a given point at a given time) as a function of S, Q, T, and a function that he termed "the Well Function." His resulting equation was

$$s = \frac{Q}{4\pi T} W\left(\frac{r^2S}{4Tt}\right) \tag{3.7–8}$$

where $W(r^2S/4Tt)$ is the Well Function.

Equation (3.7–8) furnished Theis with a graphical method for determining the formation constants by measuring in observation wells nearby the rate and magnitude of the decline of head caused by pumping water from a well at a rate Q. The Theis' method has been extensively discussed in the literature.[1,16,17]

Jacob[16] has simplified the Theis'method by noting that after some time has elapsed, and for a given distance from the pumping well, the Theis' equation above reduces to

$$s = \frac{2.30Q}{4\pi T} \log \frac{2.25Tt}{r^2S} \tag{3.7–9}$$

Therefore, to find S and T (the formation constants) using Jacob's modification of the Theis' method one plots drawdown versus time in semilogarithmic paper. The resulting graph should be a straight line. Then, one calculates S and T from the graph using the formulas

$$T = 264Q/\Delta s \tag{3.7–10}$$

and

$$S = 0.30Tt_0/r^2 \tag{3.7–11}$$

where

T = transmissivity in gallons per day per foot,
Q = pumping rate in gallons per minute,
Δs = slope of the time drawdown graph (equal to the drawdown between any two points in time on the logarithmic scale whose ratio is 10),
S = storage coefficient,
t_0 = intercept at zero drawdown in days, and
r = separation between the pumping and observation wells, in feet.

The formation constants are the most important parameters that a hydrologist must know prior to evaluating an aquifer and projecting its capabilities as a source of groundwater supply. Not only do these quantities give us an index of groundwater motion (which is important in order to estimate travel times from the recharge area to the supply well), but they also furnish an indication of the total amount of stored groundwater.

3.8 *BASINWIDE EVALUATION*

In section 3.3 the hydrologic cycle was discussed in detail. As we then pointed out all phases of the cycle have to be in balance since the total amount of water in our planet is both fixed and finite.

Such a hydrologic balance must also exist in a given basin. This fundamental idea is the basis for the simplest method used to quantitatively determine groundwater reserves in a given basin. This method is termed *Basinwide Hydrologic Inventory*. Such an inventory in a basin leads to the following expression:

$$\text{Groundwater Reserves} = \text{Precipitation} - \text{Evapotranspiration} - \text{Runoff} + \text{Inflow from adjacent basins} - \text{Outflow into adjacent basins} - \text{Pumpage} \quad (3.8\text{–}1)$$

Thus, using data from metereological stations and well records in the basin one can arrive at a reasonable estimate of groundwater reserves.

Hydrologic inventories are most commonly used in preliminary feasibility studies. When a more accurate estimation of all water reserves in a basin is desired, one must go to a more complex polygon model. Several such models are known and they are discussed in detail elsewhere.[1] They all consist in dividing the basin into polygonal units and analyzing, using the techniques of numerical analysis, the inflow, outflow, and storage in each polygon. Then, summing the storage of all polygons one obtains the reserves of the basin in question. In addition, such polygon models have been used to predict water reserves under future conditions.

To adequately evaluate the groundwater situation in a given basin it is important to have geologic information on the basin, such as, core data, stratigraphy, and structural data. In addition, it is sometimes advantageous, as shown in the previous section, to have pumping test data.

The estimation of the water available in a basin is not the only important point that needs to be determined in order to establish an adequate water supply program. The quality of the water is just as important as the quantity. To carry a basinwide water quality investigation one needs to make a comprehensive sampling and analysis program and interpret the results using the techniques of section 3.10.

If surface waters are to be used as sources of supply in a basin, a regional evaluation of these waters must be carried out. Such a study must include:

a. analysis of stream flow records to determine whether or not storage is needed,
b. geologic study of the area,

c. detailed water quality analysis,
d. cost analysis, and
e. a study of legal rights.

Aside from evaluating the quantity and quality of the water resources in a basin, prior to proceeding to the supply stage, a cost analysis of the various alternatives must be carried out. In addition, the water laws in the area under study should be carefully scrutinized.

Most hydrologic studies are concerned with looking at problems in a regional basis rather than studying the development of the water resources of a small parcel of land. In such basinwide evaluation programs the hydrologist in charge needs geologic, metereologic, and pumping test data. In addition, both geophysical and geochemical techniques may prove very helpful.

3.9 *WATER SUPPLY*

The ultimate goal of a hydrologist is to supply his resource. The area of water supply has received great attention in recent years and has been the subject of much controversy. Even though some prophets of doom make believe that the water supply situation in the United States is very bad, this is not true. The United States has at present an ample water supply and with sound management the present favorable situation can be maintained.

Excluding saline water conversion, the 48 lower states have available to them 1200 billion gallons of water per day.[18] Of these, at present, it is economically feasible to use 600 billion gallons per day. Of these 600 we use 270 and actually consume only 60! By the year 2000, however, it will be economically feasible to use 1000 billion gallons per day. Of these we will use 800 and actually consume 180.

The seriousness of the water supply problems in the United States does not lie in an overall water shortage but in localized shortages. While the United States as a whole has an ample water supply, certain areas especially in the west and southwest do not receive much water. In these areas sound water management is of the utmost importance.

Figures 3.7 and 3.8 show respectively the runoff and the groundwater distribution in the United States. As can be seen from these figures good runoff occurs in the Atlantic Gulf Coastal Plains, and in the Pacific and Midwest. Abundant perennial groundwater supplies are found in the Atlantic and Gulf Coastal Plains, the Columbia Plateau, and the glacial sands and gravels in the East and Midwest. In addition, the Southern

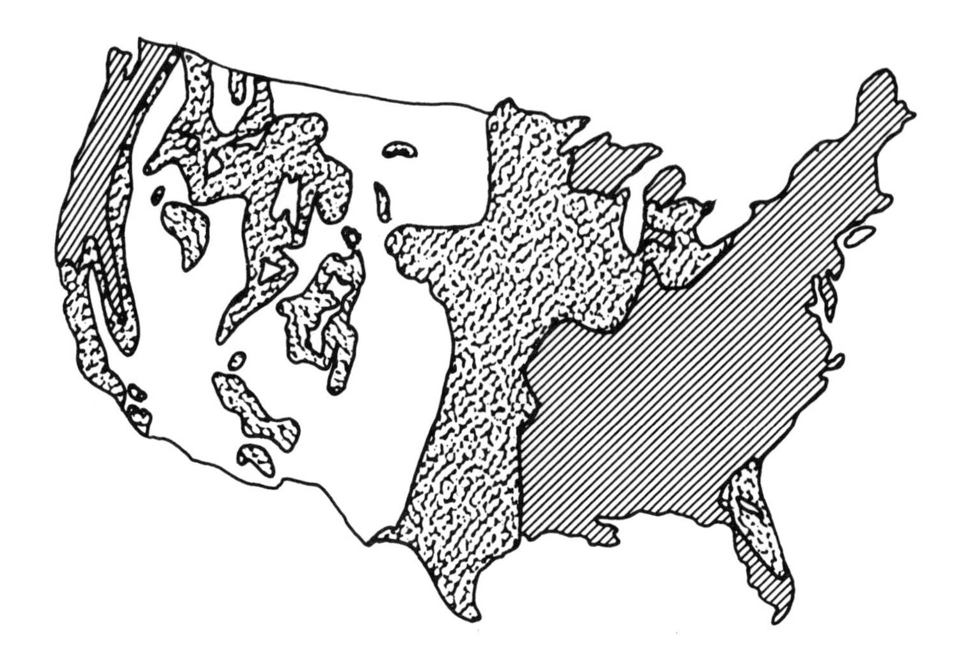

Figure 3.7. Runoff map of the U.S. (from "estimated use of water in the United States, 1960" by K. A. Mackichan, and J. C. Kammerer, 44 p. 1961, Circular 456, U.S.G.S.).

Great Plains, and the westernmost valleys of the Basin and Range Region have plentiful supplies, except, that in these areas the recharge is very low.

In order to solve future water supply problems, man is going to have to optimize to the limit his use of water. To develop adequate surface-water sources we must control the suspended and dissolved solids in the water, abate man-made pollution, and build adequate waterworks, as needed. As far as groundwater is concerned, man must stop polluting aquifers and must use only such an amount of groundwater that allows the aquifer not to be depleted.

In addition, we must refine desalination processes to make them as

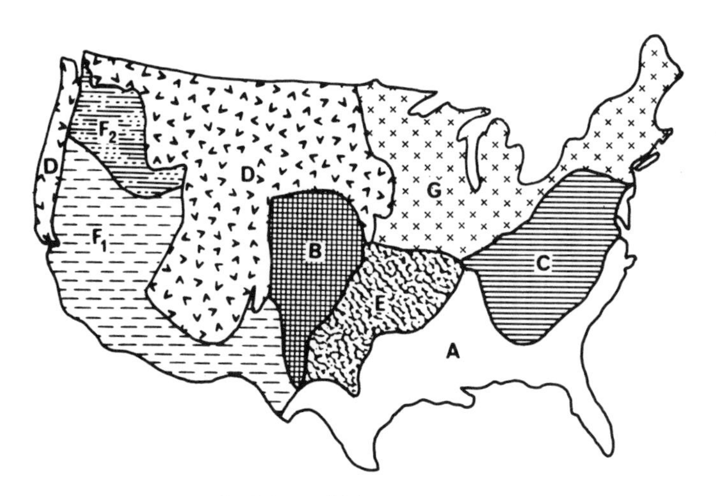

A. Atlantic and Gulf Coast Plain
abundant groundwater in sand and limestone aquifers.

B. Southern Great Plains
abundant groundwater in sand and gravel, low recharge.

C. Piedmont
limited groundwater supply.

D. Rocky Mountains, Northern Plains, Pacific Coast
small amount of groundwater, very limited potential.

E. Central Plateaus
good supplies from localized limestone and alluvium, limited potential.

F_1. Basin and Range
low recharge, good alluvial aquifers, reasonably good potential.

F_2. Columbia Plateau
good volcanic aquifers, generous recharge, good potential.

G. Glacial Zone
good sandy aquifers, bedrock productive, high recharge, excellent potential.

Figure 3.8. Groundwater in the U.S. (from "National Water Resources and Problems," select committee on national water resources, U.S. Sen., 1960.).

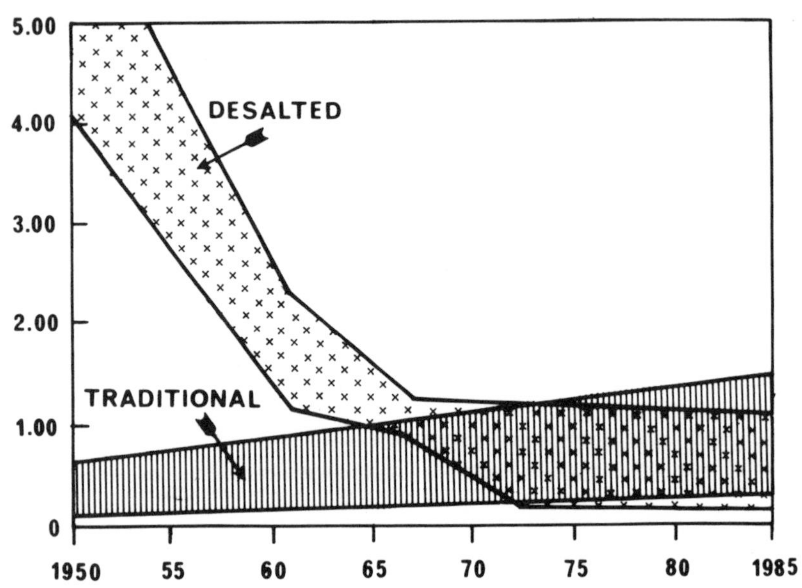

Figure 3.9. Desalting costs vs. traditional methods. (*From W.S.W. Ref. Num.* **1969** *P.R.*-**13**).

economical as possible. Recent studies by the Office of Saline Waters[19] show six possible markets for desalting processes, namely, the pulp and paper industry, the cheese industry, the iron and steel industry, the plating and metal-finishing industry, the nuclear-power-generating industry, and the treatment of acid mine drainage. The use of desalting plants offers an almost limitless source of water of utmost importance, especially, in coastal areas. The present cost of desalted water is higher than that of water obtained from conventional sources, however, the cost of desalting is going down while water from other sources is becoming increasingly costlier. By 1980 desalted water will most likely be the cheapest water available (see figure 3.9).

Finally, in order to optimize our use of water we must go into supply systems that allow almost complete recycling.

3.10 *WATER CHEMISTRY*

Up to this point we have concerned ourselves with the physical aspects of hydrology. The field of hydrochemistry or water chemistry, if you prefer, is just as important, and will be the subject of this and some of the following sections.

Water chemistry usually carries the idea of pollution in most people's minds. However, such is not the case. While pollution is certainly an important problem to a water chemist, his (her) services are also important in geochemical exploration, radioisotopic studies, and other hydrologic studies. A rather complete discussion of water chemistry can be found in more technical books on the subject.[1,20] Only certain highlights will be considered here.

Any study to describe the chemical properties of a stream or aquifer must begin by properly sampling the water body. In sampling a stream one must justify the need for each sample and try to represent, with the samples taken, the entire flow of the stream, at the sampling point, at the time of sampling. In addition, the researcher may desire to combine depth-integrated samples, equal in volume, and taken at points representing equal flow across the stream, in order to obtain an average-composition sample.

The frequency of sampling in a stream will vary depending on how rapid the water quality fluctuations are. There are streams where samples are taken three or four times a day. On the other hand, in some streams where the discharge is controlled by a reservoir, there are no visible quality changes for even one or two months.

Groundwater movement is generally slow, and not as much mixing occurs in an aquifer as with the case of surface waters. However, in any sampling program to quantitatively describe the chemical characteristics of a body of groundwater many wells must be sampled. Groundwater quality changes as water percolates through natural sediments. These changes are drastic in certain sediments, while they may be negligible in other cases.

Once the water sampling phase is completed, these samples must be analyzed for the ions desired.[20] The most important chemical constituents present in water are sodium, calcium, magnesium, potassium, silica, chloride, fluoride, nitrates, sulfates, and carbonates. Also, in many studies certain properties of the water, such as, its hardness, pH, conductivity, total dissolved solids, BOD (Biological Oxygen Demand), and temperature are reported.

Sodium is most abundant in ocean waters (8000–14000 ppm) but also occurs in connate waters, and subsurface waters that have been in contact with igneous or evaporite rocks. Calcium originates from dissolution of rocks such as limestone, calcite, aragonite, anhydrite, anorthite, and wollastonite. Magnesium is abundant in dolomitic aquifers. Potassium is usually present in very small concentrations except for thermal waters. Silica comes from the dissolution of silicates and is rarely found in excess of 200 ppm. Chloride is most abundant in sea water and usually does not exceed 200 ppm in groundwater. Fluoride is derived mostly from fluorite and appears in some volcanic gases. Sulfates are mostly derived from contact with gypsum or anhydrite. Finally, carbonates and bicarbonates are mostly the result of contact with limestone or dolomite and make the water hard.

The concentrations of the above and other ions are most commonly expressed in terms of parts per million (ppm) by weight. However, if one desires to compare concentrations of various ions present in the water one should use equivalents per million (epm). The conversion from ppm to epm can be accomplished by using the formula

$$\text{epm} = \text{ppm} \times \frac{\text{ionic charge}}{\text{atomic or molecular weight}}$$

As we pointed out in preceding paragraphs certain other properties of water are also usually reported in a geochemical study. Most important among these is pH. The pH of an aqueous solution is defined by the formula

$$\text{pH} = -\log_{10}[\text{H}^+]$$

where $[H^+]$ is the hydrogen ion concentration of the solution.

Hardness is also an important property. It is mostly due to the presence of carbonates, chlorides, nitrates, and sulfates of calcium and magnesium. Excessively hard water reduces the sudsing capacity of soap.

Hardness is commonly classified as temporary, which is due to carbonates and bicarbonates which can be removed by boiling the water, and permanent which can not be so removed. Permanent hardness is due to the presence of chlorides, sulfates, and nitrates in the water. In many chemical studies the hardness is reported as "ppm hardness as $CaCO_3$." Such values are calculated using the formula

$$\text{ppm hardness as CaCO}_3 = (\textstyle\sum \text{epm Ca} + \text{Mg} + \text{Ba}) \times 50.05$$

Conductivity is the property of a solution to conduct electricity. The larger the number of ionic constituents in a sample of water, the larger its conductivity. Conductivities are measured in the field and at the laboratory using conductivity bridges. Water conductivities are commonly expressed in micromhos (inverse microohms).

Total dissolved solids is the expression used to describe the total amount of minerals dissolved in a water sample.

Biological Oxygen Demand (BOD) is reported in sanitary or pollution analyses of water samples. It is a measurement of the amount of oxygen that is required in order to remove from the water organic matter in the process of decomposition by aerobic bacteria.

Some researchers also determine the water's COD (Chemical Oxygen Demand) which is a measure of the equivalent oxygen demand of compounds that are biologically degradable and many that are not.

The ions mentioned in the preceding paragraphs are the most prominent in ordinary water analyses. However, in some industrial pollution problems one may also be interested in other ions that may be causing problems by being discharged into a body of water. This is the subject of pollutant chemistry.

The most challenging pollution problems are posed by organic chemicals. Many organic chemicals have caused fish kills and other problems. One famous incident was the massive fish kill on the Mississippi river on November and December of 1963. This incident was the result of the insecticide Endrin getting into the water.

Table 3.2 lists some industrial pollutants and the levels that they should not exceed in natural waters, if they are to be suitable for human consumption.

As we pointed out in preceding paragraphs, another application of water

TABLE 3.2

Some industrial contaminants (data from reference 1).

Contaminant	*Suggested Maximum Level in Drinking Water (mg/lit)*
ABS	0.5
Ammonium	0.5
Arsenic	0.05
Barium	1.0
Cadmium	0.01
Copper	1.0
Cyanide	0.05
Fluoride	1.5
Lead	0.05
Mercury	0.005
Nitrates	50
Phenol	0.001
Selenium	0.01
Silver	0.05
Zinc	5.0

chemistry is in geochemical groundwater surveys. To adequately carry out such a survey a hydrologist needs to sample waters in the area and analyze for the following:

pH	Na^+	total hardness	conductivity
SO_4^{--}	K^+	SiO_2	NO_2^-
Cl^-	Li^+	B	NO_3^-
CO_3^{--}	Ca^{++}	total dissolved solids	PO_4^{-4}
HCO_3^-	Mg^{++}	total acidity	water temperature

The objective of such a survey is to elucidate the origin of groundwater in a basin, determine the chemical nature of aquifer rocks, and approximate the nature of the underground flow regime.

In order to achieve the above objectives an interpretative procedure devised at the beginning of the century by Palmer and revised later by others can be used.[1] For this purpose one begins by performing the following calculations (in epm):

$$\text{Sum of Free Acids} = \sum SO_4^{--} + Cl^- + NO_3^- + NO_2^- = FA$$
$$\text{Sum of Weak Acids} = \sum CO_3^{--} + HCO_3^- + PO_4^{-4} = WA$$
$$\text{Sum of Alkalies} = \sum Na^+ + K^+ + Li^+ + NH_4^+ = A$$
$$\text{Sum of Earths} = \sum Ca^{++} + Mg^{++} = E$$

On the basis of the four sums calculated above, one can classify the waters according to Palmer's system. This classification is as follows:

Class I	$FA < A$
Class II	$FA = A$
Class III	$FA > A$ and $FA < A + E$
Class IV	$FA = A + E$
Class V	$FA > A + E$

Once the class has been determined one can proceed to calculate the salinities and alkalinities. These quantities are defined as follows:

S1: The primary salinity in a water is the amount of alkalies that are balanced by free acids, that is, the amount of salts resulting from the union of alkalies with free acids.

S2: The secondary salinity is the amount of alkaline earths balanced by free acids.

S3: The tertiary salinity is the excess of free acids over alkalies and alkaline earths. Waters where $S3 > 0$ have free acidity.

A1: The primary alkalinity in a water is the amount of alkalies that are balanced by weak acids.

A2: The secondary alkalinity is the amount of alkaline earths that are balanced by weak acids.

A3: The tertiary alkalinity is that due to the presence of heavy metals. Most waters have $A3 = 0$. If $A3 > 0$ some nearby mineralization must be present. The common heavy metals causing tertiary alkalinity are copper, lead, zinc, iron, and manganese. The letter M will denote the sum of heavy metals.

The values of the salinities and alkalinities vary depending on the class (Palmer) of the water. They can be calculated according to the following rules:

Class I

$$S1 = 2FA$$
$$A1 = 2(A - FA)$$
$$A2 = 2E$$
$$A3 = 2M$$

Class II

$$S1 = 2FA$$
$$A2 = 2E$$
$$A3 = 2M$$

Class III

$$S1 = 2A$$
$$S2 = 2(FA - A)$$
$$A2 = 2(A + E - FA)$$
$$A3 = 2M$$

Class IV

$$S1 = 2A$$
$$S2 = 2E$$
$$A3 = 2M$$

Class V

$$S1 = 2A$$
$$S2 = 2E$$
$$S3 = 2(FA - A + E)$$
$$A3 = 2(M - S3/2)$$

The salinities and alkalinities can be used to explain certain characteristics of the water. If the primary alkalinity is the predominant value, then the water must have percolated through sedimentary alkali beds. If the secondary alkalinity predominates, the water percolates through sedimentary strata rich in alkaline earths. Usually waters where A2 is predominant possess temporary hardness. When A3 predominates, it implies that the water passed through mineralized strata. Such waters possess free acidity.

If the primary salinity is dominant, it implies that the water passed through volcanic terrain. If S2 predominates the water is usually one that has passed through sedimentary strata, volcanic in origin. Generally, such waters have permanent hardness. Finally, if S3 predominates, the water must have percolated through mineralized terrains of volcanic origin. Such waters usually possess free acidity.

Some other tools available to a hydrologist carrying out a regional water chemistry survey are the geochemical index, basin-concentration maps, semilogarithmic plots, and triangular diagrams. These techniques are discussed in detail in more technical books on the subject.[1]

Actually, the tools of water chemistry have not been used to their fullest extent by hydrologists. This field presents an excellent opportunity to newcomers for challenging work.

3.11 *WATER TREATMENT AND REUSE*

The rapid increase in water usage in most countries of the world during the past few years has led to a tremendous increase in the amount of wastewater. Such waters should be reused whenever possible and for most cases this involves some treatment procedure.

Wastewaters can be reclaimed in various ways[1] depending on the utilization to be given to the reclaimed water. In any such project, the following five objectives should be kept in mind:

i. optimization of water usage,
ii. minimization of waste,
iii. prevention of nuisance conditions,
iv. protection of health, and
v. conservation of natural resources.

TABLE 3.3.

Some water treatment techniques.

Primary Treatment
Screening
Skimming
Sedimentation
Chlorination
Clarification
Clarification-Flocculation
Secondary Treatment
Biological Oxidation
Chlorination
Digestion
Aeration
Tertiary Treatment
Soil Spreading
Flocculation
Distillation
Freezing
Solvent Extraction
Electrodialysis
Ion Exchange
Adsorption
Foaming
Chemical Oxidation

There are numerous techniques which can be used to treat wastewaters. Some of these are listed in Table 3.3. Treatments are classified as to primary, secondary, and tertiary (or advanced) according to the stage when they are applied. Figure 3.10 shows the Grand Island Nebraska Combined Municipal and Industrial Wastes Plant.[21] This plant with a design flow of 5.8 million gallons per day contains two 90′ diameter (Dorr-Oliver) primary clarifiers, two 75′ diameter aeration tanks (secondary), and two 100′ diameter final clarifiers. Each aeration tank is equipped with four dual impeller aerators with air addition to the bottom impellers.

Clarification is a common procedure both as an initial and final step in water treatment. It consists in the removal by settling of all particles in suspension that are heavier than water. Its main objectives are the removal of both flocculated and precipitated impurities so as to make the water less hard, more colorless, and less turbid. The theory behind the design of clarification tanks is beyond the scope of this book. Interested readers can refer to the work of Fair and Geyer.[22]

Another important water treatment technique is aeration. The object of this process is the addition of oxygen to the water and the maintenance of aerobic conditions. Aeration is possibly the oldest and certainly, at present, the most economical approach to BOD removal. There are many types of aerators, and while some only aerate at the surface, others combine both surface and subsurface aeration. Figure 3.11 shows a picture of a typical surface aeration unit (photograph courtesy of Dorr-Oliver Inc.).

Tertiary or advanced treatment is the last step in the purification scheme. Deju[1] and Middleton[23] discuss tertiary treatment techniques in detail. Adsorbents, for example, have been effectively used in numerous installations. Ion exchange and distillation are also popular methods. Final clarification is used sometimes as a tertiary treatment, for example, at the Grand Island Nebraska Plant previously mentioned. Solvent extraction is a most interesting type of tertiary treatment.[1] This process is simply an organic separation step followed by air stripping of the clean water effluent.

Regardless of the techniques used, the objectives of wastewater treatment are the optimization of water usage and the minimization of waste. In the very near future, managers of water resources are going to have to be concerned with more and more reuse of water. They are going to have to break the psychological barrier that surrounds in most people's minds the utilization of reclaimed wastewater.

Reuse of water is not a new idea. It has been extensively practiced for many years by numerous industries. However, the direct reuse of externally-fed wastewater in a given industry will depend on how competitive reuse is with a fresh-water source.

Refineries and petrochemical plants usually pretreat municipal effluents

Figure 3.10. The Grand Island Nebraska combined municipal and industrial wastes plant (*photograph courtesy of Dorr-Oliver Inc.*).

Figure 3.11. Photograph of a surface aerator (*courtesy of Dorr-Oliver Inc.*).

TABLE 3.4.

Recycling of water in industry. Data taken from "Water in Industry" (1965) by The National Association of Manufacturers and The United States Chamber of Commerce.

Industry	*Average Number of Times a Gallon of Water is Recycled*
Automotive	2.62
Beet Sugar	1.48
Coal Preparation	14.91
Corn and Wheat milling	1.22
Distilling	1.51
Food Processing	1.19
Machinery	5.50
Meat	4.03
Petroleum	7.62
Pulp and Paper	3.02
Soaps and Detergents	3.08
Steel	1.60
Tanning	1.04
Textiles	1.30
Natural Gas Transmission	2.32

prior to their reuse. In many occasions these industries produce condensed steam out of wastewater to satisfy their process water needs.[24]

Table 3.4 shows the extent of water recycling in industry. The coal and petroleum industry have extensively decreased their water costs by relying on recycling to a great extent.

Eller, Ford, and Gloyna[24] have done some work on the economics of recycling water. They have studied and compared in detail various recycling schemes. They summarize their views on the economics of recycling in the following five points:

"1. The feasibility of recycling with respect to cost depends upon the savings appreciated by handling a smaller volume of water as compared to the cost of treating the effluent for recycle.
2. Recycling systems are best when contaminant additions are either low or easily removed and quality requirements not stringent. Cooling operations, for example, have low water quality requirements and the heat contaminant is easily removed.
3. Recycle systems often allow for product recovery not possible in once-through systems.
4. Recycle systems offer more attraction as final effluent requirements become more stringent.

5. Once-through systems are necessary when the flexibility of quality requirements for water use is constrained, which is true in food processing and brewing operations."

3.12 *SERIOUS PROBLEMS IN WATER CONSERVATION*

As anyone who watches television or reads the newspaper well knows, there are many water pollution problems that must be solved. It is, however, difficult for the average individual to distinguish between "real" water pollution problems and those surrounded by "mass hysteria." The object of this section will be to shed some light on some of the most basic water pollution problems being presently tackled. Our discussion here will be limited to facts and conclusions based on them. Readers should avoid making far-fetched speculations on the basis of the data presented here.

The Foaming of Detergents

The word "detergent" means different things to different individuals. By detergent we will mean a cleaning compound which contains as part of its make-up some synthetically derived surface active agents (surfactants).

The problem with detergents was first detected in 1947 in Mount Penn, Pennsylvania where a heavy foam blanket was seen on the aeration tanks of the sewage treatment plant.[25] The constituent responsible for this foaming was alkyl benzene sulfonate (ABS). Residues of ABS foam at extremely low concentrations (less than 1 ppm) and have caused many unsightly foaming incidents. Nevertheless, ABS residues are not toxic and no fish kills have occurred from ABS-laden residues.

ABS was replaced in the United States in 1965 by LAS (linear alkyl sulfonate). The linear alkyl chains do not have the high degree of branching present in ABS and therefore are more biodegradable.

At present, a third generation detergent is available using olefin sulfonates. These substances are made from long-chain alpha olefins and produce a detergent that is almost 100% naturally biodegradable.

Thus, we can conclude that the problem of detergent foaming has been presently solved reasonably well.

Phosphates in Water and Eutrophication

In addition to surfactants, detergents also contain phosphates. The purpose of these is to provide softening, adjust the pH of the wash, and enhance the overall cleaning ability of the detergent.

When phosphate-loaded wastes are added to bodies of water there is a noticeable increase in eutrophication (advanced biological aging of the body of water) as a result of an excessive growth of algae. This is a very serious problem which eventually leads, if unchecked, to the death of the body of water.

Not all scientists agree that phosphates are the key to eutrophication. There are some that believe that carbon and not phosphate is the controlling factor in the process of algal growth.[26] The differences and similarities between the two schools of thought are shown on Table 3.5. All evidence so far indicates that in all probability a reduction of phosphates in detergents[27] will be effective in somewhat controlling algal growth in many of our surface water bodies.

The most promising substitute for phosphates, up to now, has been NTA (Na-nitrilo-triacetate), however, recently reported research results of a study program sponsored by the United States Department of Health, Education, and Welfare have shown that NTA can cause birth defects in animals when combined with metallic wastes. Thus, manufacturers are now staying away from this product.

TABLE 3.5.

Different views on eutrophication (taken form reference 26). (Permission granted by American Chemical Society)

Carbon-Is-Key School Believes	*Phosphorus-Is-Key School Believes*
Carbon controls algal growth.	Phosphorus controls algal growth.
Phosphorus is recycled again and again during and after each bloom.	Recycling is inefficient: some of the phosphorus is lost in bottom sediment.
Phosphorus in sediment is a vast reservoir always available to stimulate growth.	Sediments are sinks for phosphorus not sources.
Massive blooms can occur even when dissolved phosphorus concentration is low.	Phosphorus concentrations are low during massive blooms because phosphorus is in algal cells, not water.
When large supplies of CO_2 and bicarbonate are present, very small amounts of phosphorus cause growth.	No matter how much CO_2 is present, a certain minimum amount of phosphorus is needed for growth.
CO_2 supplied by the bacterial decomposition of organic matter is the key source of carbon for algal growth.	CO_2 produced by bacteria may be used in algal growth, but main supply is from dissociation of bicarbonates.
By and large, severe reduction in phosphorus discharges will not result in reduced algal growth.	Reduction in phosphorus discharges will materially curtail algal growth.

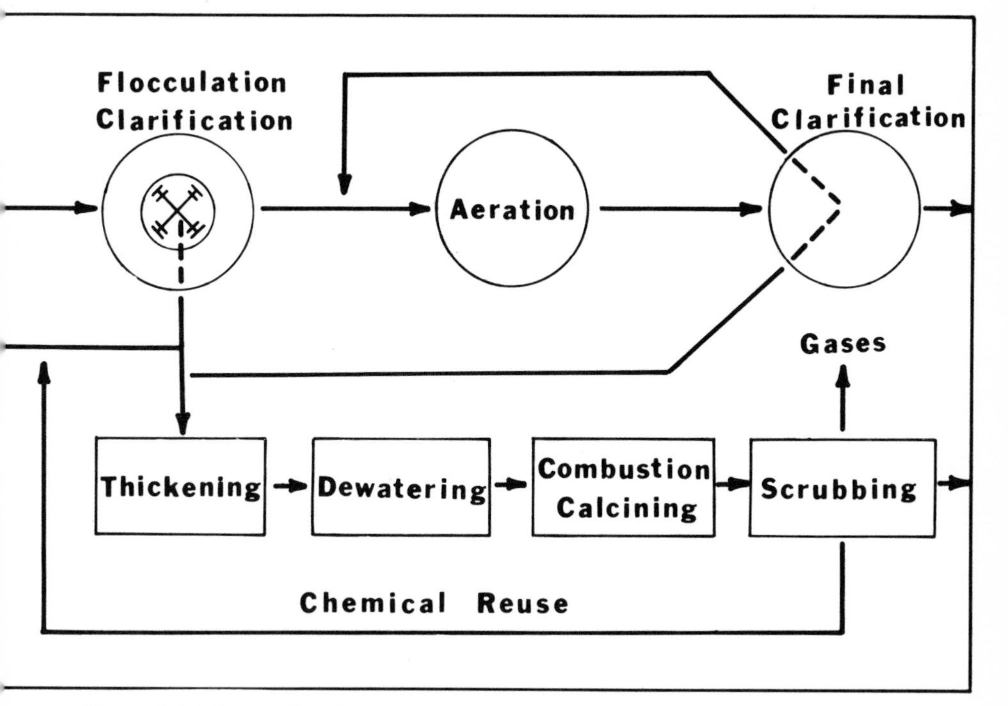

Figure 3.12. Schematic of a process removing phosphorus from waste effluents (*courtesy of Dorr-Oliver Inc.*).

As things stand now, it may be better, instead of seeking new substitutes for phosphates, to channel all efforts to improve sewage treatment plants so that phosphates can be removed from the wastewater prior to its discharge. This removal can be accomplished by either chemical or biological means. The chemical removal is obtained using a coagulation process. Biological removal can be accomplished by an activated sludge system.[28,29]

Integrated sewage systems that also remove phosphorus are at present under operation in some parts of the country. Figure 3.12 shows a schematic of a process designed by Dorr-Oliver Inc.[30] for the removal of phosphorus.

In contrast with the problem of detergent foaming, the problem of eutrophication is one that has not been solved yet and one that will most probably be the subject of much heated debate in the next few years.

Mercury Pollution

A very complex water pollution problem is caused by the discharge of mercury into our bodies of water. Mercury is used in many processes, such as, the manufacture of urethane, vinyl chloride, acetylene, batteries, etc. Table 3.6 shows how extensive the consumption of mercury was in the United States during 1969. Among these users, many discharge some mercury as effluent waste. Prominent among these is the chloro-alkali industry. During 1969 this industry discharged into the environment an average of 0.30 lbs of mercury per ton of caustic soda produced.

Mercury is very detrimental to human and animal health. Its effects can eventually lead to death. Initial symptoms of mercury poisoning are tremor, irritability, hearing and visual deterioration, difficulties in walking, talking, and swallowing, tingling, and numbness of the fingers. In later stages, the subject is involuntarily mobilized, can suffer nearly total blindness, and his mental functions rapidly deteriorate until death arrives.

Mercury poisoning is not a new phenomenon. In fact, the expression "mad as a hatter" resulted from the symptoms that hatters developed as a result of poisoning from mercuric nitrate used in the processing of felt.

TABLE 3.6.

Mercury consumption in the United States during 1969 (data taken from Chemical and Engineering News, June 22, 1970).*

Electrolytic chlorine	1,572,000 lbs
Electrical apparatus	1,382,000
Paint	739,000
Instruments	391,000
Catalysts	221,000
Dental preparations	209,000
Agriculture	204,000
General laboratory use	126,000
Pharmaceuticals	52,000
Pulp and paper making	42,000
Amalgamation	15,000
Other uses	1,082,000
	6,035,000 lbs

* Figures were compiled by Chemical and Engineering News from data supplied by U.S. Bureau of Mines.

An early incident involving mercury pollution of waters occurred in 1952 at Minamata Bay, Kyushu, Japan. This incident was the result of dumping mercury-loaded wastes into the bay. The incident caused the death of 44 persons and crippled 70 others. In addition, many cats who ate fish from the bay went completely berserk.

In the United States, the Food and Drug Administration (FDA) has set a limit of 0.5 ppm mercury in fish to be used for human consumption. Fish with concentrations 10 or 15 times higher than this limit have been caught in the Great Lakes and in other areas.

There are several forms of mercury and their poisoning capacity is different. Methyl and alkyl mercury are very poisonous because they cause irreversible damage to the central nervous system. Elemental and phenyl mercury only produce reversible damage to the kidneys and intestines. However, phenyl and inorganic mercury, when released into the water, can undergo methylation and be converted into methyl mercury. The methylation process occurs very rapidly in winter and early spring, and preferably in waters of low pH, and under aerobic conditions.[31]

To restore a mercury-contaminated lake to a clean state several approaches are presently possible. However, of these, some are inefficient, others are too expensive, and some substitute one evil by one which may perhaps be worse. It is of the utmost importance to presently reduce mercury discharges to the lowest possible minimum until we develop a more comprehensive picture of this problem and its possible solutions.

Pollution in Flotation Operations

A common technique used in the recovery of valuable minerals is the process of froth flotation. The job of controlling the extent of pollution from a flotation operation should begin during pilot-laboratory testing. The engineers designing a given flotation operation should carefully consider not only the reagent costs, concentrate grade, and percent recovery, but also, the pollution problems inherent to each flotation scheme.

The pollution problems involved in flotation stem mostly from the rather noxious reagents used in the process, such as ferrocyanide, ferricyanide, xanthates, amines, dithiophosphates, and sulfonates. Therefore, one must avoid seepage of flotation effluents into the soil, accidental spills, and breakage of the tailing dams, in order to prevent serious water pollution incidents.

In addition, flotation plant effluents must be effectively treated prior to their discharge.[32] Some of the treatment methods commonly used are:

(a) chemical alteration and precipitation,
(b) coagulation and sedimentation,

(c) activated charcoal adsorption,
(d) biological treatment,
(e) filtration, and
(f) foaming.

Some installations use chlorine oxidation to remove xanthates, cyanide, sulfides, and dithiophosphates, while others prefer carbon adsorption,[33] and still others use suitably acclimated growths of microorganisms. Regardless of the treatment procedure, great care must be taken at all times in order to avoid the introduction of noxious flotation reagents into usable sources of water.

Mine Acid Drainage

A most important water pollution problem is the drainage of mine acid into the adjacent lakes and streams. This is a very serious problem especially in the coal mining regions of the northeastern United States. The acid that forms and then drains into the streams is the result of the reaction between pyrite, oxygen and the water that filters into an operating or abandoned mine. The pyrite is usually found in the mine tailings or in and between coal seams. The reaction by which the acid forms is:

$$2FeS_2 + 7O_2 + 2H_2O \rightarrow 2H_2SO_4 + 2FeSO_4$$

The extent of mine acid drainage is tremendously large and stems both from abandoned and working mines. In Pennsylvania alone, over 1600 tons of acid ($H_2SO_4 + FeSO_4$) are dumped daily into 2500 miles of the state's streams.

There are five basic methods of treating mine acid pollution. These are:

1. lime or limestone neutralization,
2. chemical treatment, recycling, and recovery,
3. closing the mine,
4. sealing all abandoned sections of a working mine, and
5. water diversion so as to keep the mine dry.

The most popular of these techniques are lime and limestone neutralization. These processes are as follows:

$$\text{(lime)}\ CaO + H_2SO_4 \rightarrow CaSO_4\downarrow + H_2O$$

$$\text{(limestone)}\ CaCO_3 + H_2SO_4 \rightarrow CaSO_4\downarrow + H_2O + CO_2\uparrow$$

Mine acid can render a stream useless and kill all its animal life. It is important that discharges of mine effluents always be on the alkaline side (pH 7–9) and that their iron content be low (less than 7.0 mg/liter).

Steel Industry Pollution

Another important problem is the disposal of spent pickle liquor used in the steel industry. This liquor is an aqueous solution that contains about 5% acid and 10% iron. In the United States alone, roughly a half billion gallons per year of spent liquor is produced. At present, even though regeneration and recovery techniques are available, they are not favored since disposal is the easiest and cheapest thing to do. However, it is very probable that in the near future steel executives will be tempted to use regeneration and recovery techniques.

Oil Pollution

The oil industry has also had its share of water pollution problems. The most spectacular of these have been oil spills in the high seas and well blowouts in offshore areas. In addition the oil industry has had to be concerned with other water pollution problems due to the discharge of dangerous refinery effluents into streams and to the discharge of hydrocarbons as a result of pipeline leakage.

To avoid all these problems numerous safety precautions must be taken in all phases of exploration, drilling, production, and transportation of hydrocarbons. In addition, research to improve cleanup techniques in the high seas, harbors, and beaches should be undertaken.

Pesticide Pollution

The problem of pesticide pollution has recently received much heated debate. Ecologists have recently intensified their efforts to reduce or completely curtail the use of pesticides such as DDT.

DDT is the common name for a chlorinated hydrocarbon insecticide known formally as 2,2-bis(*p*-chlorophenyl)-1,1,1-trichloroethane. DDT has been remarkably effective as an insect killer. It has a remarkable persistence and its killing effects can last as much as 2 years. It is relatively insensitive to sunlight and its solubility in water is small. However, its effectiveness as an insect killer has decreased because many insects, such as the common fly, have developed an immunity to DDT.

The water pollution problems associated with DDT and other pesticides are of two types:

(1) groundwater contamination due to the percolation of water through pesticide-sprayed soil, and
(2) surface water contamination due to runoff through pesticide-sprayed soil.

Various researchers have reported examples of pesticide pollution. For example, Butler[34] reported that DDT adsorbed by Atlantic Coast oysters interferred with the marine life in the area. Moats and Moats[35] have traced the loss of lake trout in a New York State Hatchery to DDT residues in eggs. Other researchers[36] discuss other aspects of pesticide pollution and the effects of such pollution. Nevertheless, it should be clearly noted that so far no human deaths other than those due to accidents can be attributed to pesticide pollution.

What is needed to solve the pesticide problem is a pesticide that is effective against a given pest and then self-destructs itself after a short period of time. Experiments have been carried out to develop a self destructing form of DDT[37] by combining it with a zinc catalyst. However, much more work is yet needed.

Thermal Pollution

In the generation of electrical power, especially by nuclear plants, much heat is given off into adjacent bodies of water. This thermal pollution tends to accelerate fish metabolism, speed up respiration in fish, and affect their appetite, rate of digestion, and growth. If the change in temperature is very abrupt it can cause the death of fish since they are cold-blooded creatures that cannot control adequately their body temperatures.

However, the thermal discharges from a power plant may be used advantageously by man in any of the following ways:

1. to irrigate crops and warm the soil,
2. to help maintain an aquaculture of certain species during the winter months,
3. to prevent water freeze-up in streams during the winter months,
4. as a heat source, and
5. by injecting the hot water deep in the ocean to cause upwelling of the deeper waters (below the thermocline) and raise nutrients into the upper waters thus helping in the development of nearby fisheries.

For the case of thermal pollution we have a discharge which can either be regarded as a source of pollution or as a source of revenue. If this dual nature of other residues were analyzed many pollution problems would be solved.

Summary of Water Pollution Problems

The problems discussed above illustrate the complexity and variety of water pollution problems that one must be able to solve. Research by industry, government, and university laboratories has significantly contributed to the solution of many water pollution problems and will continue to solve, or at least ease, many others. Such research, coupled with what we can learn from past mistakes, and with sound management should help in creating an improved water environment for present and future generations to enjoy.

3.13 *THE ROLE OF WATER FOR INDUSTRY*

It should be clear to the reader by now that water resources play a most important role in the development of man's activities. Abundance of cheap water can make a community prosper, and lack of it, can cause a speedy death. Industries select their locations partly on the basis of abundance of reasonably priced water and, in many instances, they are also interested in sites where disposal of wastes can be easily handled.

Industrial growth in the United States has been tremendous. Similarly, use of water by industries has grown at a fantastic pace, much greater than the level of production. Many authorities now estimate that in 1980 the industrial use of water could be as much as 400% above 1955 figures.[38]

Industries planning to relocate or build a new plant in a given site usually carry out a thorough hydrologic study to determine whether water is going to be a blessing or a problem at the site under consideration. Such hydrologic reconnaissance must include a study of the quantity and quality of the water available, be it, surface water or groundwater. In addition, the fluctuations in quality and quantity as a function of climatic conditions should be investigated. Other points to be considered are the problem of water rights, water costs, waste disposal sites, possibilities for expansion, need and/or advantages of recycling, etc.

In some cases it may be advantageous for a given industry to develop its own source of water and its own water treatment facility. This point should be considered in the preliminary reconnaissance. If it is shown that it is better to have an independent system, then, tests should be undertaken to determine what interference, if any, will occur between the new system and present users. For example, an industry planning to tap a given aquifer may need to drill deeper than all present users in order not to interfere with them.

Much can be achieved by such planning, and industry can, in the long run, save many dollars and headaches by so doing. In most instances, communities and local chambers of commerce are very cooperative and lend a hand to the industrialist who wants to undertake such planning.

3.14 *FUTURE OUTLOOK*

In the preceding sections of this chapter we have looked at the field of hydrology from many angles. It was pointed out earlier in the chapter (see section 3.9) that, even though, we should not be ready to press the panic button as far as the water situation is concerned, we should be busy working to optimize water utilization, preserve its quality, and prevent water shortages. All this can be accomplished by sound water management and with community support. The key to the solution of problems in the area of water pollution is not to sit and wait, but to solve the problems before they happen or at worst before they lead to irreversible consequences.

As far as water supply is concerned, some areas in the United States are likely to suffer water shortages in the next decade, unless, adequate measures are taken to meet the projected demands. The projections for the 80's and 90's are not gloomy but they are not 100% attractive. For the next decade the northeastern United States will have adequate water supplies with the exception of the Delaware-Hudson area which will need to import water.[38] Some other, rather localized, regions in the northeast will need by 1980 additional water storage facilities. The southeastern United States and the Midwest will most likely have adequate amounts of water in the next decade. The area of the Rockies and the Basin and Range Province (see Figure 3.8) will reach their maximum supply level in the 80's and will thereafter need to import water. The Pacific Northwest and the Columbia Plateau have ample water for years to come, however, in the Pacific coast, southern California will be importing water within the next decade.

In general terms, it is very probable that the entire concept of water supply will undergo re-shaping within the coming decade. Desalting will become more attractive, even though, extensive utilization of desalting will probably not take place due to the problem of transporting the water. Recovery of water from brines and geothermal areas will also be done in the next few decades. However, this will only have localized importance. Most important, industries and municipalities will move, more and more in the next few years, into using complete or partial recycling schemes.

In the area of water pollution, regulations for disposal of effluents are bound to become more stringent. This will increase most manufacturing costs which simply implies higher costs for consumer goods.

The environmental control battle will be one of the prime objectives of the United States Government for the next decade. Taxpayers will bear the cost of such programs. What are all such programs going to lead us to? Hopefully, they will help preserve the quality of our waters, and guarantee that we will have enough water to meet the nation's demands in the years to come.

3.15 *REFERENCES*

1. Deju, R. A. (1971) "Regional Hydrology Fundamentals," Gordon and Breach Science Publishers Inc., New York, N. Y.
2. Babbitt, H. E., Doland, J. J., and Cleasby, J. L. (1962) "Water Supply Engineering," McGraw Hill Book Co. Inc., New York, N. Y.
3. Walton, W. C. (1970) "Groundwater Resource Evaluation," McGraw Hill Book Co. Inc., New York, N. Y.
4. Johnson, E. E., Inc. (1966) "Ground Water and Wells," E. E. Johnson, Inc., St. Paul, Minnesota.
5. Jacobs, J. A., Russell, R. D., and Wilson, J. T. (1959) "Physics and Geology," McGraw Hill Book Co. Inc., New York, N. Y.
6. Baker, Jr., R. A. (1969) *Design and Operation of Large Desalting Plants*, Water and Sewage Works Reference Number, pp. 4–16.
7. McIlhenny, W. F. (1968) *Chemicals From Sea Water*, Proceedings of the First Conference on Materials Technology, ASME, pp. 119–126.
8. The Dow Chemical Company (1967) "A Feasibility Study on the Utilization of Waste Brines From Desalination Plants, Part I," Office of Saline Water Research and Development, Report No. 245.
9. Mero, J. L. (1965) "The Mineral Resources of the Sea," Elsevier Publishing Co., Amsterdam.
10. DeWiest, R. J. M. (1965) "Geohydrology," John Wiley and Sons, New York, N. Y.
11. Meyer, A. F. (1942) "Evaporation From Lakes and Reservoirs," Minnesota Resources Commission, St. Paul, Minnesota.
12. Mesnier, G. N., and Iseri, K. T. (1963) "Selected Techniques in Water Resources Investigations," U.S.G.S.-Water Supply Paper #1669-Z.
13. Darcy, H. (1856) "Les Fontaines Publiques de la Ville de Dijon," V. Dalmont, Paris.
14. Krumbein, W. C. and Monk, G. D. (1943) *Permeability as a Function of the Size Parameters of Unconsolidated Sand*, AIME Transactions, Petroleum Division, volume 151, pp. 153–163.
15. Terzaghi, K. (1943) "Theoretical Soil Mechanics," John Wiley and Sons, New York, N. Y.

16. Jacob, C. E. (1950) *Flow of Ground Water*, in Engineering Hydraulics, chapter 5, Hunter Rouse (ed.), John Wiley and Sons, New York, N. Y.
17. Theis, C. V. (1935) *The Relation Between the Lowering of the Piezometric Surface and the Rate and Duration of Discharge of a Well Using Ground Water Storage*, Transactions American Geophysical Union, volume 16, p. 520.
18. Guyton, W. F. (1963) *Planning the Plant Water Supply*, Chemical Engineering, June 10, pp. 170–174.
19. Editor (1970) *Desalters Eye Industrial Market*, Environmental Science and Technology, volume 4, p. 634.
20. Rainwater, F. H., and Thatcher, L. L. (1960) *Methods for Collection and Analysis of Water Samples*, U.S.G.S.-Water Supply Paper #1454.
21. Dorr-Oliver Inc. (1969) "ABC System for Primary and Secondary Waste Treatment," Dorr-Oliver Inc., Stamford, Conn., 14 pp.
22. Fair, G. M., and Geyer, J. C. (1958) "Elements of Water Supply and Waste-Water Disposal," John Wiley and Sons, New York, N. Y.
23. Middleton, F. M. (1970) *Advanced Treatment of Wastewaters for Re-Use*, in Water Re-Use. . . The Objective of Advanced Waste Treatment Research, Scranton Gillete Publications, pp. 18–27.
24. Eller, J., Ford, D. L., and Gloyna, E. F. (1970) *Water Reuse and Recycling in Industry*, Journal of the American Water Works Assoc., volume 62, pp. 149–154.
25. Brenner, T. E. (1969) *Biodegradable Detergents and Water Pollution*, Advances in Environmental Science, volume 1, pp. 147–196.
26. Editor (1970) *The Great Phosphorus Controversy*, Environmental Science and Technology, volume 4, pp. 725–726.
27. Sawyer, C. N. (1952) *Some New Aspects of Phosphates in Relation to Lake Fertilization*, Sewage and Industrial Wastes, volume 24, pp. 768–776.
28. Levin, G. V. (1963) *Reducing Secondary Effluent Phosphorus Concentration*, First Progress Report, Dept. of San. Eng. and Water Res., John Hopkins University.
29. Levin, G. V., and Shapiro, J. (1965) *Metabolic of Phosphorus by Wastewater Organisms*, Journal Water Pollution Control Fed., volume 37, pp. 800–821.
30. Albertson, O. E., and Sherwood, R. J. (1967) *Phosphate Extraction Process*, paper presented at Pacific Northwest Section, Water Pollution Control Fed., Oct. 25–27, 1967, reprinted 1968 by Dorr-Oliver Inc., 22 pp.

31. Editor (1970) *Mercury in the Environment*, Environmental Science and Technology, volume 4, pp. 890–892.
32. Davis, F. T. (1970) *The Control of Water Pollutants From Flotation Plants*, paper presented at the Pacific Southwest Mineral Industry Conference, 20 pp.
33. Demidon, V., and Volodin, V. (1946) *Sorption Methods of Removing Cyanide Compounds From the Wastewater of Concentration Plants*, Tvensky Metaly.
34. Butler, P. A. (1969) *Monitoring Pesticide Pollution*, Science, volume 19, pp. 889–891.
35. Moats, S. A., and Moats, W. A. (1970) *Toward a Safer Use of Pesticides*, Bioscience, volume 20, pp. 459–463.
36. Macek, K. J. (1968) *Reproduction in Brook Trout—Federal Sublethal Concentrations of DDT*, Journal of Fisheries Research Board of Canada, volume 25, pp. 1787–1796.
37. Anonymous (1970) *Pesticides*: *Self-Destructing DDT*, Chemical and Engineering News, volume 48, pp. 12–13.
38. Anonymous (1968) "Water's Role in Plant Location," Water Information Center Inc., Special Report, Long Island, New York, 8 pp.

Chapter 4

THE OCEANS

4.1 INTRODUCTION

An ocean is basically a dilute sodium chloride solution plus all the other soluble constituents that have been discharged into it. Its chemical composition is fairly uniform except where the water circulation is restricted or near where constituents are being added or removed. Near the mouths of rivers the salinity of the ocean decreases while the content of silt, organic matter and any other dissolved constituents the river may be carrying increases. The salinity of the oceans' surface waters increases near the equator because the heat from the sun evaporates large quantities of water while leaving the salt behind. The wind sometimes pushes this warm, high-salinity water into high latitudes where the water is cooled. The cooling increases the waters' density such that some of it slowly sinks to form the "deep water" of the ocean. This circulation is slow, but mixing does take place, and a fairly uniform ocean composition is maintained.

Ocean life has adjusted to the comparatively constant environment of ocean water. The rapid changes in temperature common to the atmosphere that are tolerated by terrestial life often cannot be tolerated by sea life. Some species may cease to grow, live, or reproduce while conditions are not tolerable. Other species which were marginally adapted to the old environment may increase their population. This more adapted species may change the entire ecology of the area.

The life cycle of the ocean depends upon the phytoplankton (passively-floating plant life) living in the euphotic zone where photosynthesis can occur.[1] The euphotic zone will vary with latitude, season, degree of turbidity, etc. Light may penetrate the water to 100 meters (330 feet) depth in the tropics where the water is clear and light rays approach the vertical over a considerable part of the day. In the temperate regions the euphotic zone may be less than 50 meters (160 feet) deep in the summer and much shallower in the winter. Phytoplankton take in carbon dioxide and give off oxygen as land plants do. The concentration of phytoplankton in ocean water is less than that of plants on land such that the net energy used for photosynthesis on land is slightly higher than that used in the ocean even though there is much more sea surface than land surface. The low concentration of phytoplankton is primarily due to low nutrient concentrations in the euphotic zone and high metabolism. Yet, the many varieties of phytoplankton synthesize more than 90 percent of the basic material that feed the life of the sea.[2] Herbivorous zooplankton and some small fish graze on the phytoplankton. Various carnivorous creatures live by eating the grazing creatures, and they are later eaten by other carnivorous creatures both large and small. Finally, the debris from the life at the surface settles to the bottom where eventually some of it re-enters the circulating water system, and is returned to the surface.

Upsetting this delicately balanced life cycle has not been advantageous to man. When certain phytoplankton become too concentrated, the water turns red and the fish die. Hydrogen sulfide produced by decaying fish is very unpleasant. It may cause the paint on houses along the shore to peel off and make the beaches no longer desirable. When conditions become too favorable for certain starfish which eat coral, they eat so much coral that they kill the coral reefs which protect the island on which man is living. If the starfish are not stopped the islands may return to the sea. Echinoderms have begun multiplying rapidly and are destroying the kelp beds off California which have been harvested for years. In each of the above cases it has been suggested that man has specifically done something which has upset the life cycle of the ocean.

Most of the time, disruptions caused by man will be corrected by natural processes if the disruptive processes are stopped. For instance, the globs of oil now observed in the open ocean will eventually be eliminated by bacteria and other natural processes if dumping of oil at sea decreases. The present dumping rate apparently exceeds the capacity of nature to eliminate the oil.

The ocean helps to maintain the balance of gases in the atmosphere. Besides photosynthesis mentioned above, there is an exchange of gases at the surface between the ocean and the air. All types of gases are exchanged. For instance, large quantities of carbon dioxide (CO_2) created by the combustion of fossil fuels are probably absorbed by the ocean such that the concentration of CO_2 in the atmosphere has not increased as much as expected. Radioactive isotopes created by the explosion of atomic bombs are trapped by the sea, and effectively removed from the atmosphere. The sea also returns things to the atmosphere such as water and salt. The salt later serves as nuclei for the formation of rain drops (See Chapter 7), and either returns directly to the sea or first to the land then indirectly to the sea.

The wind drives the ocean and the ocean drives the wind. The ocean absorbs heat from the sun and in turn heats the air. The heat makes the wind blow and the winds move the water. Since the winds have prevailing directions they set up currents in the ocean. The winds move more rapidly than the water so they carry more heat; much of it in the form of water vapor derived from the sea. Although the water does not move as quickly as the air, it still carries large quantities of heat due to its large heat capacity and volume. The interaction between the wind, ocean, and sun are very important in determining the world's climate.

The ocean affects the weather. Cities located near the ocean have their weather modified by the temperature of the water. The ocean has a much larger heat capacity than the atmosphere. The water heats or cools the air according to the relative temperature such that cities near cold water currents are cooler than inland cities and vice-versa.

It has been suggested that a major change in the ocean currents would produce a major change in the weather. This weather change could be so great that it would cause an ice age. Long-range weather prediction is intimately related to the study of the weather over the ocean. To this purpose, weather ships and buoys are maintained world-wide to report the conditions at sea. Weather satellites have been a great advance in weather prediction since they very rapidly observe both the ocean and the land.

This chapter briefly describes the physical, chemical, and biological properties of the ocean. The relationship of estuaries to the open ocean is discussed. Man's effect on and use of the ocean is discussed with due regard for his limited knowledge of the ocean's environment and ecology.

4.2 PHYSICAL OCEANOGRAPHY

Ocean Circulation

To the casual observer ocean waters are uniform, but in truth they are not. Although the water is in motion most of the time, boundaries between different water masses can be identified, and these maintain their identity throughout the oceans. Naturally, in many cases where different water masses come together, the contact is diffuse. For example, the contact between the Gulf Stream water mass and the North Atlantic Central Water mass is visually detectable from ships as well as airplanes.[3] These water masses vary vertically as well as horizontally. Currents within the ocean can be identified by their water-type as well as their velocity. Surface waters are often different from deep waters, and thus the ocean is stratified.

Density differences of the various water masses maintain the ocean's stratification. The water changes its salinity and temperature at the surface where it can be affected by the sun and the air. The sun raises the temperature of only the first few millimeters of water because the heat energy cannot penetrate further into the water. Turbulence created by the wind mixes the thin surface layer to a depth of about 100 meters (330 ft). As the surface water temperature increases, its density decreases such that this water tends to remain at the surface and not mix below the turbulent zone. The warm water at the surface also evaporates leaving most of its salt behind. The increased salinity increases the water density tending to offset the density decrease created by the temperature rise. The range in temperatures in the ocean is greater than the range in salinity such that temperature generally determines the water density. In most cases a change of 1 ‰ (parts per thousand) salinity will produce a greater change in density than a 1°C. change in temperature. The temperature of the open ocean ranges from near −2°C. to +33°C. The salinity of the open ocean ranges from about 33 ‰ to 37 ‰.

Physical Properties

The salinity (S) of sea water[4] can be defined as the total amount of solids (in grams) present in one kilogram of ocean water after all the bromine and iodine have been replaced by chlorine, all organic matter oxidized and the carbonate converted to oxide. The value of salinity defined in this manner is slightly lower than the amount of dissolved solids in grams per kilogram. Salinity is usually measured in terms of chlorinity (Cl^-).

The number giving the chlorinity in per mille of seawater sample[5] can be defined as being the number giving the mass with unit gram of atomic weight silver just needed to precipitate the halogens in 0.3285234 kilograms of sea water sample. Chlorinity can be converted to salinity by means of the following formula defined in 1901 by Knudsen[6]:

$$S = 1.8050\,Cl + 0.03$$

This formula was redefined in 1967 to[7]:

$$S = 1.8066\,Cl$$

Modern practice is to measure the sea water conductivity compared with Standard Seawater of known chlorinity and then to calculate the salinity. Salinity and chlorinity are expressed in parts per thousand. Generally sea water has about 35 ‰ salinity and 19 ‰ chlorinity.

The density of sea water is usually a calculated value because it is very difficult to measure in-situ. Three variables are involved in calculating density, namely: salinity, temperature, and pressure. Since water is almost incompressible the effect of pressure on the density of water is negligible in most instances. The density of sea water ranges from 1.02 to 1.07 grams per cubic centimeter. The term "sigma-tee" (σ_t) is commonly used to describe density at sea level which is equivalent to neglecting pressure effects. Thus,

$$\sigma_t = (\rho - 1 \frac{g}{cc})\, 1{,}000$$

where ρ (g/cc) is the density at one atmosphere at the recorded salinity and temperature. Therefore, a density of 1.02524 g/cc becomes $\sigma_t = 25.24$ with the units given by implication.

In practice σ_t is determined by plotting salinity versus temperature on special graph paper and reading σ_t directly from the graph. With the advent of computers available at sea the manual practice will probably stop.

The Ocean as an Engine

The sun, air, and ocean interacting with each other form a large heat engine. The heat from the sun evaporates the water. Water vapor in the air reduces the heat from the sun that reaches the ocean, thus decreasing the evaporation rate. The warm ocean surface also heats the air providing the energy to make the winds blow. The winds move the water creating a turbulence which mixes the surface water with cooler, deeper water, consequently, lowering the surface temperature and reducing the heating of the air. The above short description indicates how complicated the operation of this heat engine can be. A more detailed description of this phenomenon can be found in books on meteorology and physical oceanography.[3]

The prevailing winds move the ocean in such a way that semipermanent currents have been formed (figure 4.1).[8] These currents transfer warmer, higher-salinity water from the tropics to cooler climates north and south. Although most of the heat is transferred through the atmosphere, the effect of the heat transferred by the water noticeably modifies local ocean climate.

Ocean Stratification

Ocean stratification can be detected by observing the vertical temperature variations. The surface layer of water, consisting of roughly the top one hundred meters, undergoes seasonal changes in temperature. The main thermocline layer below the surface layer is detectable by the rapid change in temperature with depth.

The main *thermocline* layer (figure 4.2) is not affected by seasonal changes. This layer commonly extends to a depth of slightly more than 1,000 meters (3,300 ft). Below the thermocline the change in temperature

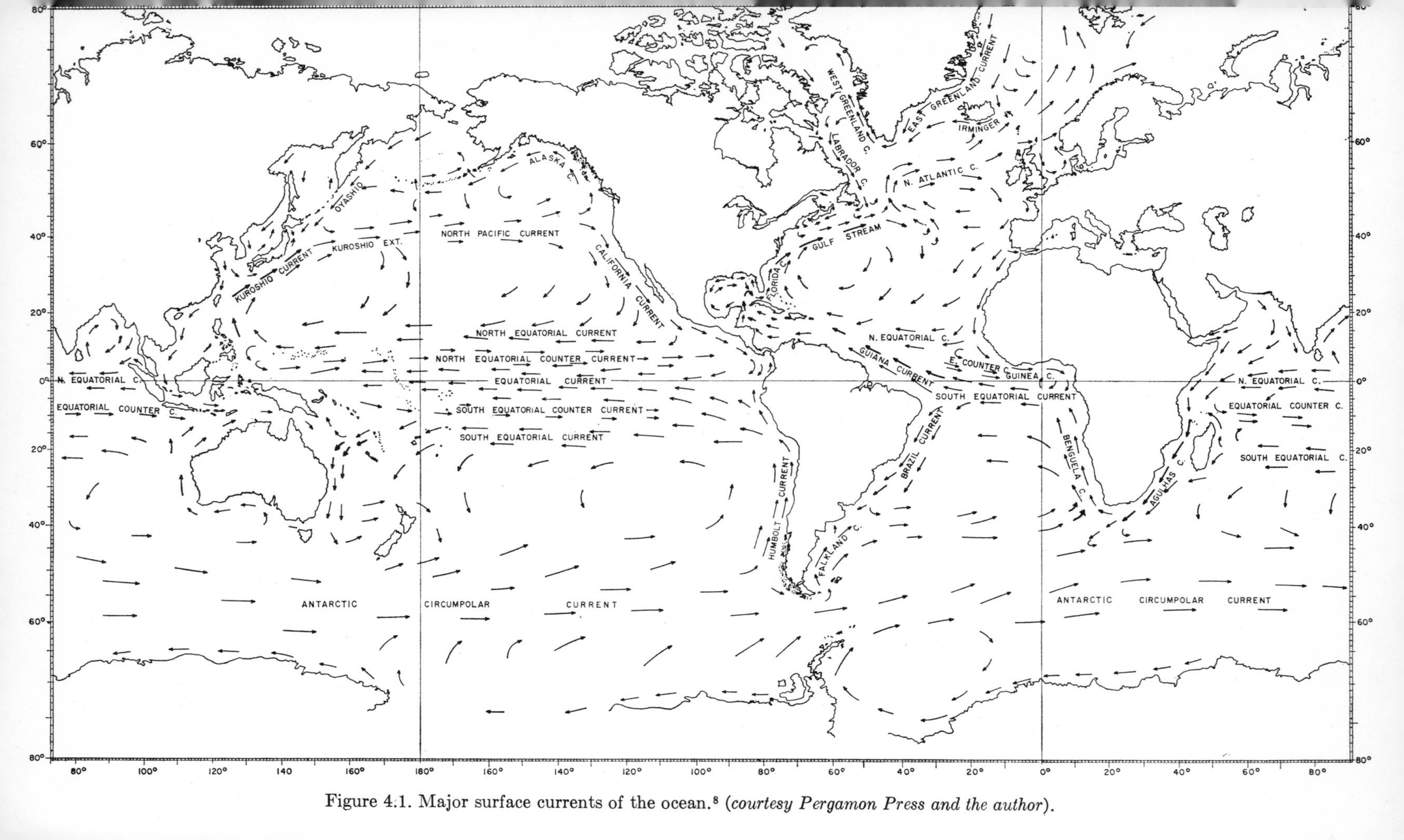

Figure 4.1. Major surface currents of the ocean.[8] (*courtesy Pergamon Press and the author*).

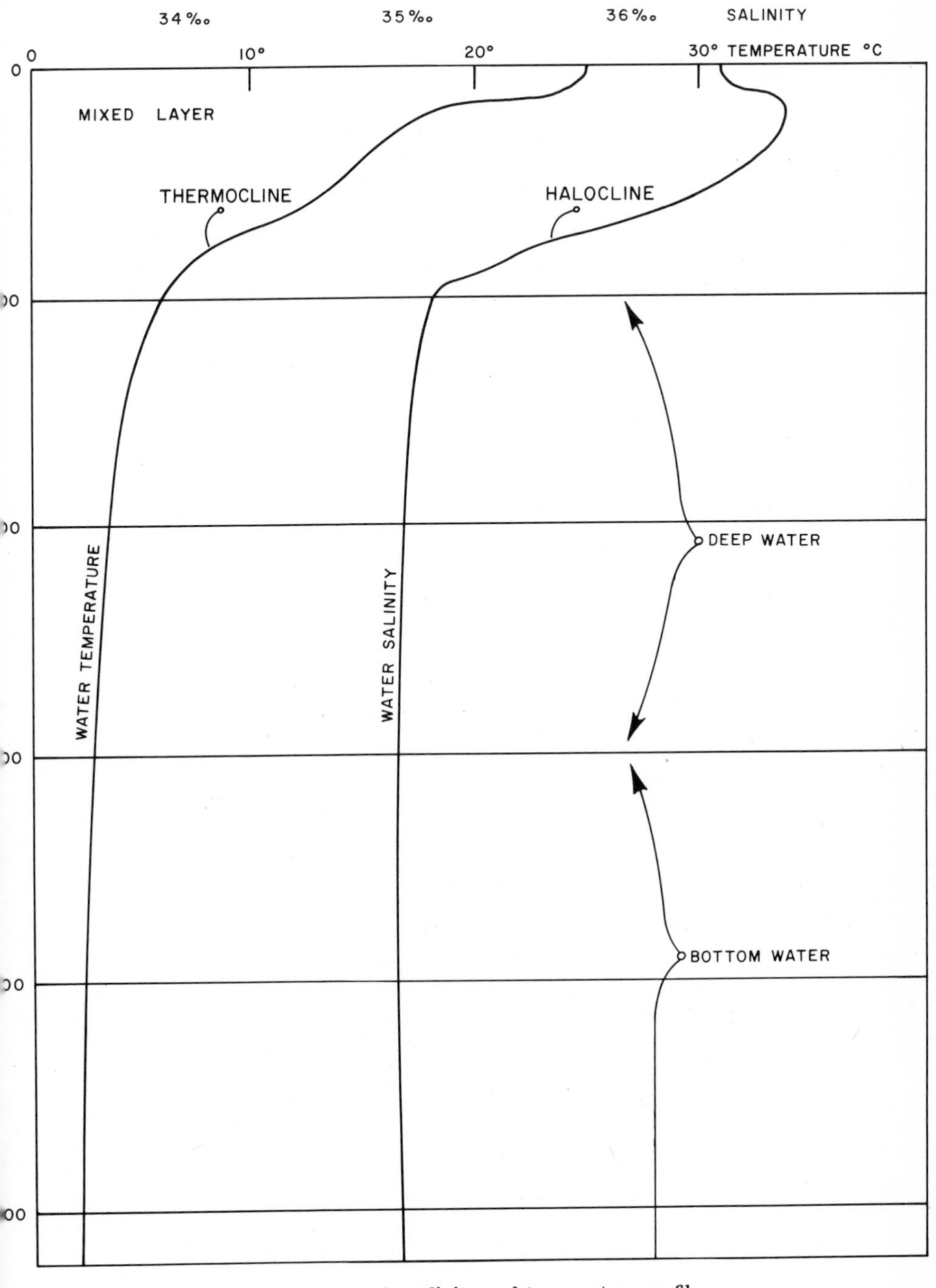

Figure 4.2. A diagram of a salinity and temperature profile.

is not so rapid. In the following 2,000 meters (6,600 ft) the temperature change of the *deep water* is approximately 3°C., whereas in the thermocline the temperature change may have been 20°C. over 1,000 meters (3,300 feet) change in depth. The *Bottom Water* lies below the deep water and includes all the water to the bottom. There is little change in temperature in this zone. The average temperature here is close to 2°C. The Bottom Water is only formed in a few places in the ocean. The Bottom Water that is formed off Greenland is more saline than the Bottom Water formed off Antarctica. In the South Atlantic Ocean the Antarctic Bottom Water lies beneath the North Atlantic Deep Water formed off Greenland (figure 4.3).[9]

Although the cold, dense Bottom Water is formed very slowly and sinks very slowly, it covers the oceans below 3,000 meters (10,000 ft). The vertical circulation of the ocean is more rapid in some locations than in others. The minimum oxygen concentration is above 1,000 meters (3,300 ft) depth and below the euphotic zone. There must be vertical circulation for oxygen to be below 1,000 meters.

Depending on the assumptions and measuring method, the age of Pacific Bottom Water is thought to be between 1,000 and 1,600 years, and Bottom Water of the Atlantic Ocean is about half that old.[7]

Tides and Waves

Tides move across the ocean in a rotational manner. The changing positions of the sun and moon with respect to the earth, and their related gravitational forces move the tidal wave. If the earth were completely covered by water of sufficient depth, the tidal wave could keep up with the moon. This wave would have a velocity close to 1,600 kph (1,000 mph). The ocean, however, is too shallow to propagate a wave of such velocity. The astronomical tide consists of forced oscillations which have periods commensurate with the tide generating forces, however amplitudes and phases are restricted by friction and water depth. Frictional forces and shallow depths cause the crest of the tide to pass sometimes many hours after the sun or moon has crossed the local meridian.

The height of the tide changes in relation to the combined gravitational attraction of the sun and moon. If the sun, moon, and earth are in a line the attraction is stronger and the tidal range is greater. As the three bodies become more out of alignment the tidal range decreases.

Prediction of actual tides is very complicated because of the several periods of oscillation related to the three celestial bodies along with the standing wave characteristics of large and small bodies of water. The net result is that prediction of tides requires numerous observations combined

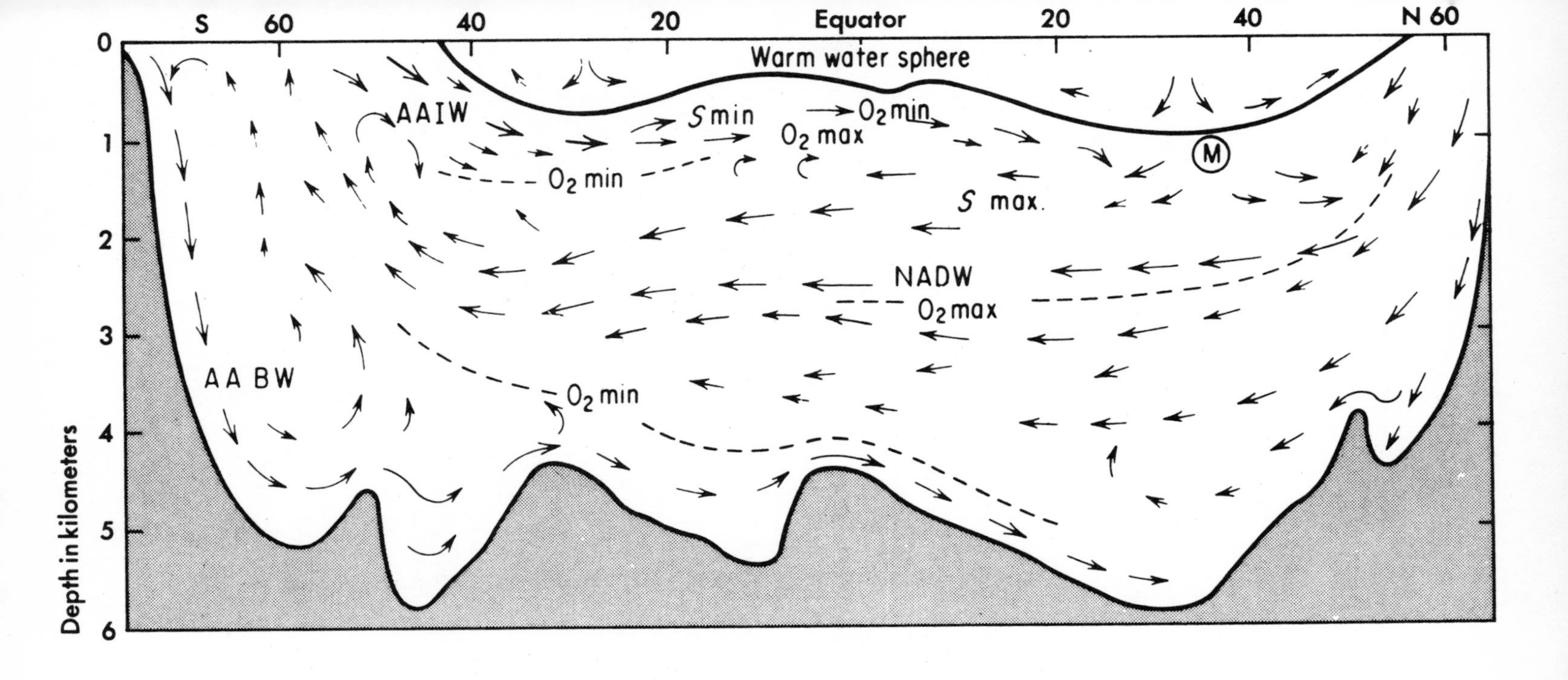

Figure 4.3. The major water masses of the Atlantic Ocean. The arrows show the main directions of water flow. AAIW is Antarctic intermediate water and M is Mediterranean water (flowing from east to west). AABW is Antarctic bottom water and NADW is north Atlantic deep water. The maximum and minimum oxygen concentrations are also shown as they help identify water masses. "S min" and "S max" are the salinity minimum and maximum layers in the deep ocean.[9] (*courtesy Prentice-Hall, Inc.*).

with mechanical or mathematical analysis. Storms affect the tides such that long-term tide prediction may not reflect the observed tidal range at a particular time.

The "tidal wave" or tsunami is not related to the astronomical tide except that it is also affected by the bottom conditions. The tsunami is a wave with a long wave length and low amplitude which moves at the maximum speed for the open ocean (about 400 knots*). It is created by a sudden movement of the ocean bottom, such as, an earthquake or volcanic eruption. The wave spreads outward from the site of the disturbance with decreasing energy. It is usually not noticeable in the open ocean, but as the wave enters shallower waters it begins to build height. When this wave enters a restricted bay, it may attain a height of over 50 feet and be very destructive (figure 4.4). Tsunamis are not necessarily single waves but may be a series of waves. The abnormally low water preceeding a crest has tempted people to walk out to observe the newly exposed bottom only to be engulfed by the incoming crest.

Ordinary waves, in contrast to tides, appreciably curve the ocean surface. The air flowing over the water disturbs it and makes the surface rough. The roughness increases the drag of the air on the water such that more energy is transferred to the water. Capillary waves, which are controlled by surface tension and have wave lengths less than 1.73 cm, are thought to be the main method of transferring the air movement to the water. Gravity waves have wave lengths longer than 1.73 cm and are controlled by gravitational and inertial forces. Wave speeds for gravity waves are decreased when the water depth is much less than one wave length. Increasing wave length for gravity waves increases their velocity. Gravity waves increase their velocity with increasing wind velocity, but capillary waves cannot do this. Eventually gravity waves will attain the velocity of the wind if the wind has sufficient fetch. When waves attain the wind velocity, the energy extracted from the wind is greatly reduced.

After the winds stop, the waves continue to move across the ocean. Different waves move at different velocities such that they interfere with each other. This interference creates longer wave lengths and therefore higher velocities. The waves being driven by the wind generally have periods of less than 10 seconds and have unsymetrical shapes. These are called *sea*. The long period waves developed after the wind stops have more symmetrical shapes and periods longer than 10 seconds. These are called *swell*. Gravity waves normally observed in the ocean are a combination of sea and swell. Symmetrical waves with different wave lengths and velocities combine to produce an irregular sea state (figure 4.5).[10] Some waves

---

* 1 knot = 1.15 statute miles per hour.

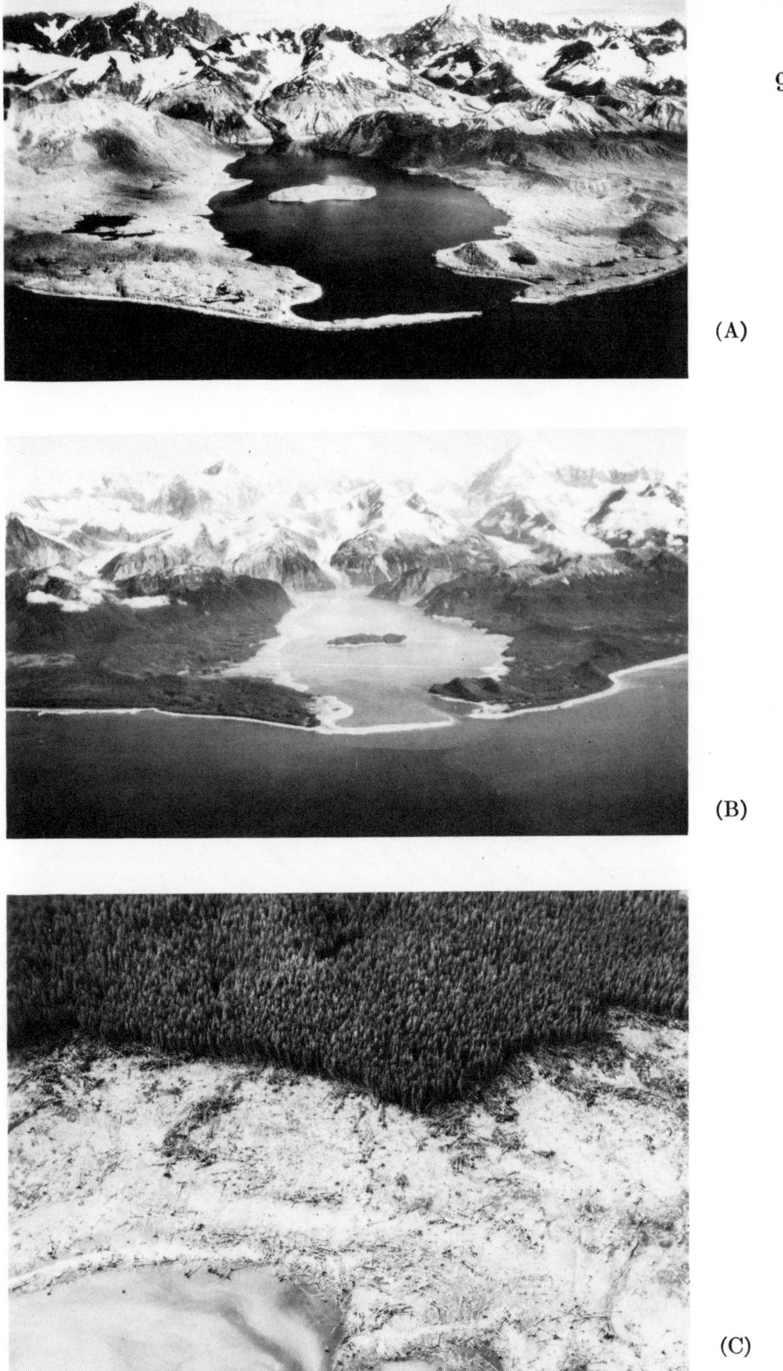

Figure 4.4. Letuya Bay, Alaska. (A) In 1954 before the giant wave. (B) In 1958 showing wave damage. (C) In 1958 showing wave damage on North Shore. (*photographs courtesy of D. J. Miller, U.S. Geological Survey*)

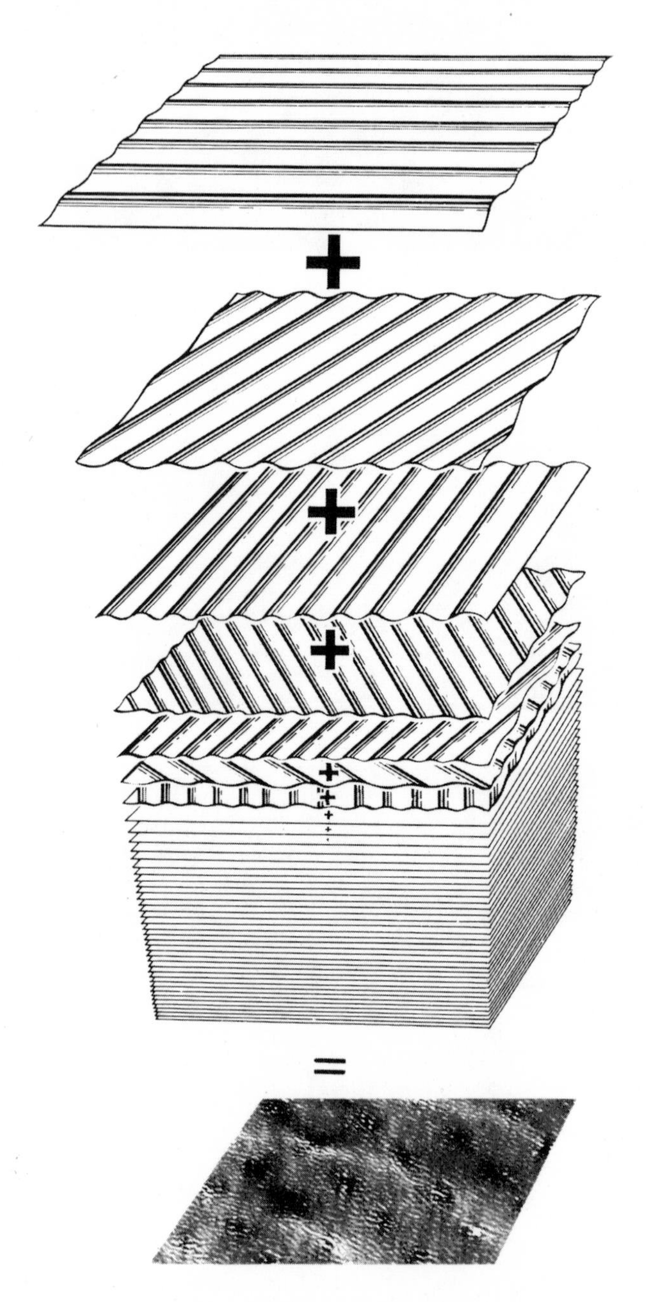

Figure 4.5. The combination of many simple waves moving in different directions produces an irregular sea state.[10] (*courtesy of U.S. Naval Oceanographic Office*)

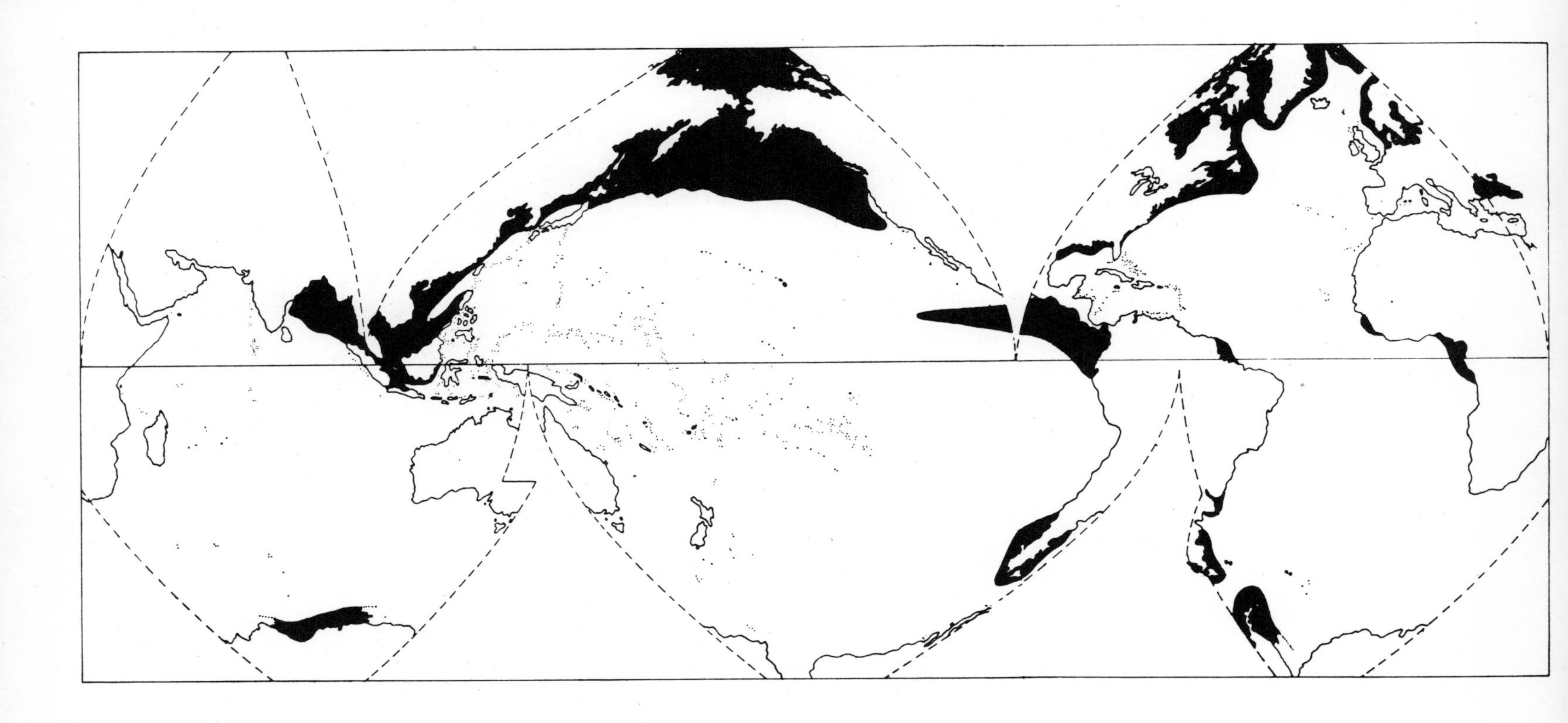

Figure 4.6. Offshore estuarine zones of the world's oceans bounded by the 33.5 ‰ isohaline.[11] (*courtesy of J. L. McHugh, copyright 1967 by The American Association for the Advancement of Science, publication no. 83, page 582*).

combine to produce a wave with a larger amplitude, whereas other waves combine to produce another wave having an amplitude less than either of the original waves.

Estuaries

Pritchard's[11] definition of estuary includes all semi-enclosed coastal bodies of water that possess a free connection with the open sea and within which sea water is measurably diluted with fresh water coming from land drainage. McHugh[11] finds Pritchard's definition too limiting for fish because fish swim great distances away from the Pritchard estuary, but are dependent for part of their life cycle upon the Pritchard estuary. McHugh uses the second part of Pritchard's definition to include all the water between the 33.5 ‰ isohaline and land (figure 4.6).[11] The water within the 33.5 ‰ isohaline is less saline than the open ocean and therefore more related to the estuary. In some areas the source of the diluting fresh water can be determined to come from a particular estuary such as the Mississippi River. McHugh's estuary includes much water which has oceanic circulation, whereas Pritchard's estuary has unique circulation.

Within the confines of an estuary formed at the mouth of a river the salinity varies from that of the fresh water flowing down the river to that of the open sea. McHugh would include all the water out to the 33.5 ‰ isohaline, but Pritchard would stop much closer to shore where the water circulation becomes oceanic. Estuarine circulation is quite variable yet some patterns can be seen. The interaction between the flow of the tides moving saline water, and the river moving fresh water creates currents which vary with the seasons, weather, and tides. The fresh water is less dense than the saline water and flows over it. The boundary surface between the two types of water varies with the velocity and volume of each type of water. The shape of the shore and the bottom of the bay modify the current flow. The layered structure of the water body may be modified depending upon the intersection of the water boundary with the confining land. Under storm conditions the range of salinities may be completely changed to either oceanic or river type. Under calm weather and steady river flow the water body may become stratified such that the bottom waters are depleted of oxygen, anaerobic conditions are created, and bacteria produce toxic hydrogen sulfide.

Fresh river water flowing out to sea tends to bend toward the right (in the northern hemisphere) of the direction in which it is flowing, and therefore keeps to one side of the estuary. The inflowing tidal water also bends to the right such that it keeps to the opposite side of the estuary from the outflowing river water. When the tide reverses, the oceanic water tends to flow on the same side of the estuary as the river water. Thus, the oceanic water has a slightly counterclockwise current motion. This is reflected in the boundary surface between the fresh and the saline water. The boundary rises and falls according to the direction of the tide. The higher part of the boundary surface moves across the estuary to reflect the greater volume of flow.

All ocean currents tend to the right in the northern hemisphere and to the left in the southern hemisphere because of the Coriolis force which is an apparent force due to the earth's rotation.

4.3 BIOLOGICAL OCEANOGRAPHY

Plants and animals are fundamentally different. Plants can produce their own food because they have chlorophyll. Animals do not have chlorophyll, and therefore must eat plants or other animals to obtain their food.

Chlorophyll is a group of green catalytic pigments essential for photosynthesis. In photosynthesis light energy is absorbed, and simple materials are chemically reduced and converted to more complex substances.[12] The chlorophyll within phytoplankton (free floating plants) absorbs light energy present in the euphotic zone, thereby converting the sun's energy to chemical energy. This chemical energy is then available to promote other biochemical endothermic reactions by which the plant is constructed.

An example of the photosynthetic process is the production of glucose and oxygen by the following reaction:

$$\underset{\text{Carbon dioxide}}{6CO_2} + \underset{\text{Water}}{6H_2O} \xrightarrow[\text{Chlorophyll}]{\text{Light}} \underset{\text{Glucose}}{C_6H_{12}O_6} + \underset{\text{Oxygen}}{6O_2}$$

This process requires 690 K cal/mole of glucose which in the ocean is supplied by the sun's radiant energy. Thus, the process must occur in the euphotic zone (figure 4.7)[12] which normally limits it to the upper 80 meters (260 feet) of ocean. As the light intensity diminishes the rate at which photosynthesis occurs decreases. Plants also have respiratory chemical reactions which consume oxygen, and are not related to light intensity. Therefore, with diminishing light the depth of compensation may be reached wherein the amount of oxygen produced is equal to the oxygen consumed. Below this depth the respiratory processes dominate, and plants are no longer productive (figure 4.8). A plant that gets into this zone eventually dies if it is not returned to a zone of more light. If the

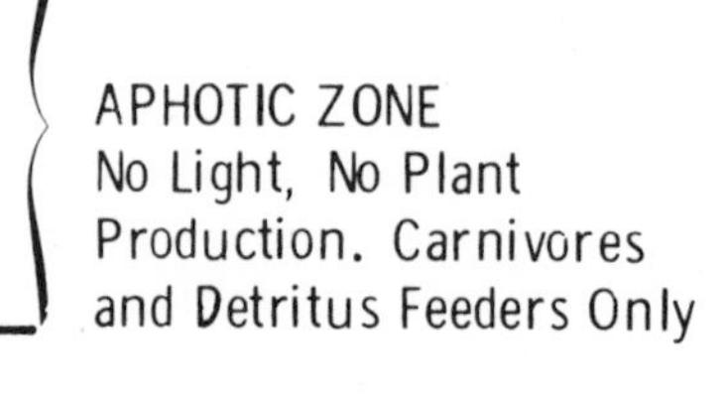

Figure 4.7. Penetration of light into the sea.[12] (*courtesy of Wiley-Interscience*)

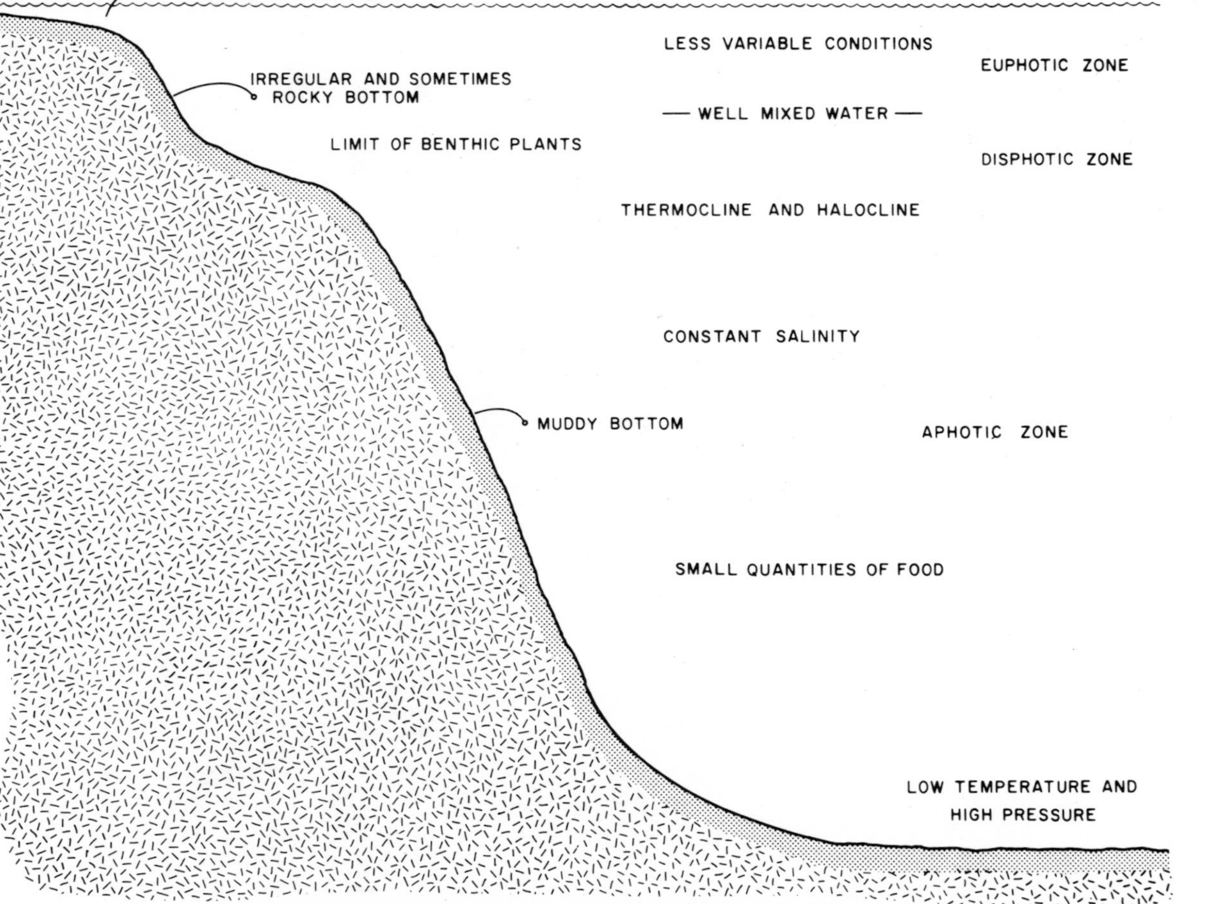

Figure 4.8. The relative location of some conditions within the marine environment.

plant dies, respiratory processes will eventually degrade the organic compounds built by photosynthesis and return the extracted nutrients to the sea.

Plants have developed several ways to enhance their ability to remain in the euphotic zone. The most common adaptation is to increase their surface area, and thus the frictional resistance to water. The increased surface area also brings the plant in contact with a larger amount of nutrients. Other flotation adaptations include special shell shapes, thin shells, and secreted oils or fats that lower the bulk density of the plant. In addition, fats store food which may be consumed by grazing animals such as zooplankton (free floating animals) (figure 4.9) and fish.

The rate of photosynthesis is apparently inhibited by too much light as indicated by the rate decreasing in the upper few meters of the ocean. Carbon dioxide (CO_2) is required for photosynthesis, but is usually in abundant supply in the ocean, and is therefore not a limiting factor.

Some essential nutrients such as sulfur, potassium, and magnesium are usually found in the ocean in sufficient concentrations for photosynthesis to occur. Nitrogen, phosphorus, and iron are found in lesser quantities, and sometimes their lower concentrations limit plant growth. The large surface area of ocean plants aids in acquiring the necessary nutrients by absorbing them from large volumes of water.

Nutrient distribution in the ocean is not uniform. Some seasonal changes are related to nutrient use by phytoplankton during periods of extensive growth and reproduction. The phytoplankton consume the available nutrients until starvation limits reproduction and growth. Geographical differences which affect phytoplankton growth also affect nutrient concentrations. In high latitudes phytoplankton production is relatively low beneath the ice cover which limits the penetration of light. Where there is sufficient light, and the water circulation brings large quantities of nutrients to the surface, high phytoplankton production does not deplete the nutrient supply.

Even though the obvious nutrients are present in ocean water, the light is sufficient, and the temperature and salinity are acceptable; sometimes the phytoplankton production decreases. Production is therefore limited by some other factors which are not properly understood at present. Many trace elements are present in the plants which may be necessary for growth. Vitamins may also be important. Copper, zinc, and manganese are known to help growth. Silicon is required for the shells of diatoms, and when the silicon concentration is low the diatom shells become thinner.

Salinity influences the osmotic pressure between the environment and the organism. The salinity limits the type of organism which can grow in a certain environment.

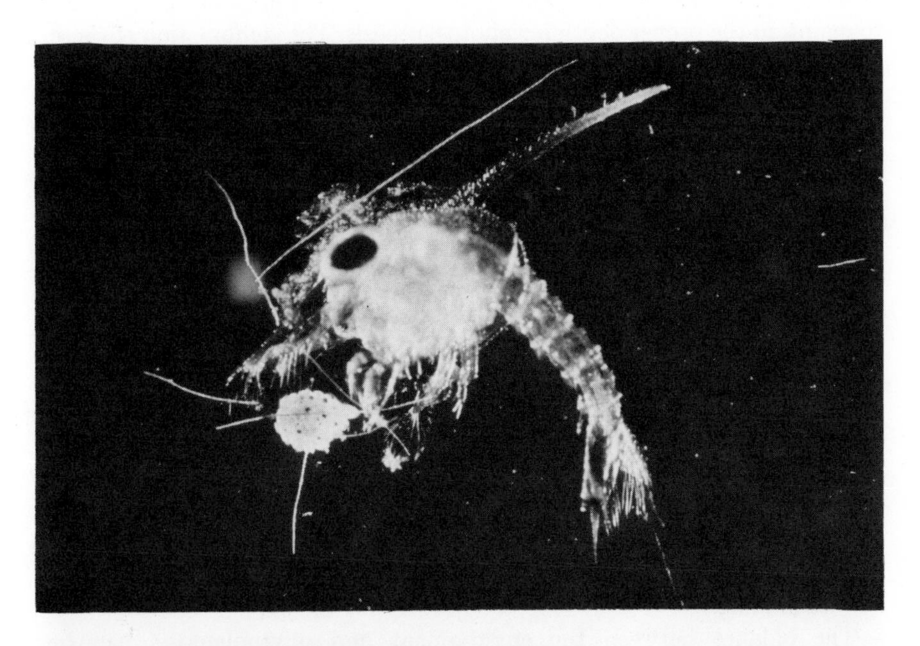

Figure 4.9. Examples of Marine Plankton. The size of Plankton is indicated by the coin (a dime) in the lower photograph. (*photos by Moyer-Stormont Undersea Media*)

Temperature affects the metabolism of organisms. It also affects water viscosity and density which will in turn affect the ability of an organism to maintain the proper depth for photosynthesis. The combination of these factors affects the distribution of the organisms.

Light has more effect on the seasonal changes in phytoplankton production than temperature. Production generally peaks in the spring in temperate and high latitudes. Rapid decrease in production is usually accompanied by a decrease in nutrients. A second smaller peak generally occurs in the autumn.

It has been suggested that availability of light, and a stable upper layer of water cause the spring bloom of phytoplankton. Too much deep mixing may remove the phytoplankton from the euphotic zone. The autumn bloom occurs after storms replenish the nutrients in the upper layer by disrupting the thermocline such that nutrients can again reach the surface. The autumn bloom is smaller than the spring bloom because there is usually less light available.

Where vertical mixing regularly brings nutrient-rich deeper waters to the surface, the ocean may support a large phytoplankton population. Mixing may occur due to upwelling, divergence, turbulence, and convection. The very productive areas in figure 4.10[7,13] are areas of more constant vertical mixing.

The balance between the environment and phytoplankton can be changed by changing the limiting factors. Raising the temperature increases the metabolism of some plants and may be beneficial in some situations. In other situations the plants may die. Increasing the concentrations of certain nutrients under the correct conditions may produce a bloom which is beneficial and will increase the food available throughout the entire food chain. Efforts to increase vertical mixing in the open ocean are aimed at increasing the general food supply by bringing nutrients from below the euphotic zone into the euphotic zone. Where conditions of salinity and temperature create a stable stratified body of water, without vertical circulation, the oxygen may be depleted by the phytoplankton which fall below the euphotic zone; thereby producing anaerobic conditions followed by the bacterial production of hydrogen sulfide. Oxygen starvation or hydrogen sulfide may then kill animals in the anaerobic zone. If vertical circulation occurs rapidly, the hydrogen sulfide may reach the surface waters and kill life there.

"Red tides" are the result of phytoplankton blooms which due to the high density of organisms turn the water a reddish color.[1] Some of these organisms produce toxins which kill the fish. Thus, the food supply may be temporarily decreased by the increase of nutrients in the water.

Outfalls from sewage disposal plants along the California Coast have

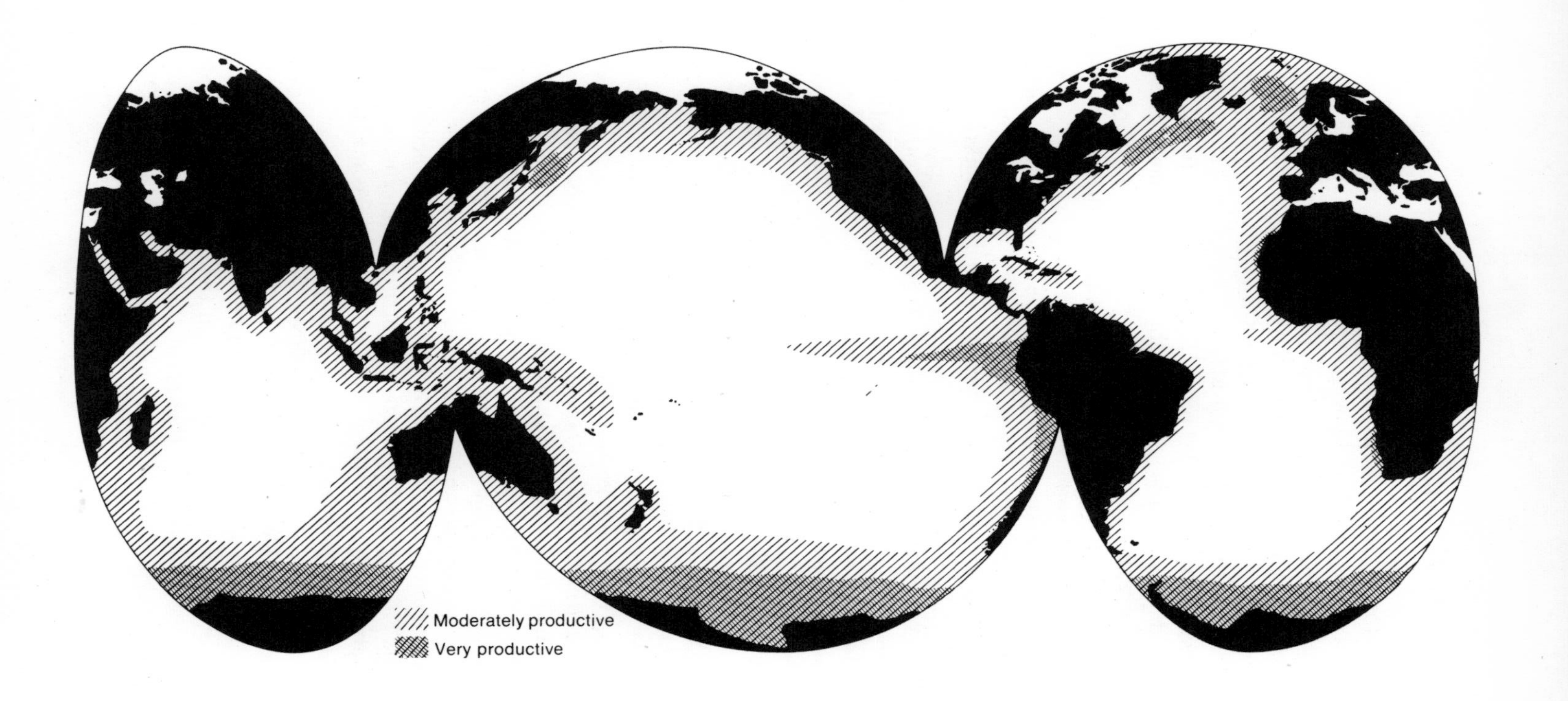

Figure 4.10. Probable plant productivity of areas within the ocean. Light gray indicates low productive areas.[7, 13] (*reproduced by permission of Appleton-Century-Crofts, Education Division, Meredith Corporation, and the Sears Foundation for Marine Research*)

increased the phytoplankton concentration to the extent that the beaches are littered when on-shore winds blow the plants toward shore. On the other hand, dumping of waste along the Atlantic seaboard has created some areas almost devoid of sea life. A third example of the effect of waste disposal is dumping of automobiles and tires into the sea to create artificial reefs where benthic animals can attach themselves and prosper.

Not all plants which live in the ocean are planktonic (free-floating). Some are attached to solid objects within the euphotic zone. Others are attached for part of their lives and unattached for other parts. Animals also may go through planktonic stages, and later attach themselves to the bottom or become free swimming.

Zooplankton graze upon the phytoplankton, and obtain the chemical energy and nutrients which are stored in the latter. If the supply of nutrients has limited the phytoplankton growth, then the zooplankton growth is also limited.

This limiting of animal and plant growth continues upward through the chain of animals which feed upon each other. The big fish eat the little fish. Some of the plants and animals concentrate unnecessary chemicals in their fatty tissues. These chemicals may have little effect upon the plants, but as they are passed up the food chain the chemicals may eventually encounter animals to which the chemicals are toxic. For example, the brown pelican is possibly facing extinction because of DDT it obtains from eating fish. The ingested DDT causes the pelican egg shells to be so thin that they often are broken before the birds are ready to hatch.[14]

The benthos are the organisms which live in close proximity to the bottom of the sea. They may go through planktonic stages which permit them to be spread over the bottom before they attach themselves to some suitable object. Some of the benthic animals crawl along the bottom while others burrow into the sediment. Plants form part of the benthos in the euphotic zone. Bacteria of the benthos may form an important part of the food chain in the deep ocean.

Generally, the number of individuals living on the bottom decreases with increasing depth. This is probably related to the availability of food which must fall from above. Many of these animals are filter-feeders, but they must have organic material to eat.

The nekton are the free swimming animals which can move through the water independent of the current. No plants are included in this group. These animals can search actively for food and avoid other free swimming predators. They are capable of migrating when the environment becomes uninhabitable. Some members of this group have very selective areas, such as estuaries for reproduction from which they cannot migrate. When the spawning areas become unsuitable, these animals may cease to reproduce.

Most nekton feed on plankton. Of course, the larger nekton feed on the smaller ones. Nekton can limit and control the plankton population. When the nekton die, they become the food for the benthos of the deep sea if they are not devoured by something else as they slowly fall through the water to the bottom.

Estuaries are very important in the biological cycle. They are natural nutrient traps. They receive much of the waste material and sediment from the land. The nutrient value of the waste material may be quite high and beneficial, or it may be toxic to the marine life of the estuary. Too much sediment may bury the benthos or disrupt their reproduction. Too many nutrients may produce plankton growth which may deplete the oxygen in the water such that nekton and benthos may die due to oxygen starvation. Some sort of balance needs to be maintained.

As mentioned earlier, the nutrient concentrations in the euphotic zone of the open ocean often limit the quantity of flora and fauna which may live in a particular area. Estuaries are natural traps for nutrients as well as the route nutrients must travel from land sources to the sea. In order to prevent eutrophication and/or poisoning the estuaries, efforts are being made to prevent waste material from reaching the estuaries. Some nutrients in this waste, which could be utilized by sea life, are almost permanently removed from the food cycle.

Two definitions of estuaries were mentioned earlier. The nektonic estuary of McHugh[11] (figure 4.6) is bounded by a salinity front defined at 33.5 ‰. Within this estuary many nekton, particularly fish, return to the Pritchard estuary at some time in their life cycle. In effect some of these fish, which have migrated hundreds of miles, never left the estuary since they remain within the boundaries deliniated by a particular range of salinities.

Fish are well adapted to the estuarine environment. Their scales, skin and coat of mucus minimize the biological stress of their system caused by changes in osmotic pressures associated with changes in salinity. Their swimming ability also permits them to avoid unfavorable conditions. However the young of many marine species, in which these protective systems apparently are not well developed, migrate to water of very low salinity in estuaries. When they become older, they migrate to more saline waters.

There is something in the environment of the estuary which is necessary for the growth of the nekton. If the environment of the estuary is changed chemically or physically such that the water is not acceptable for breeding at the proper season, the effect on the oceanic nekton population will be quite noticeable.

The benthos of the estuary must be quite tolerant to changes in salinity and temperature. Many of the benthos have developed adaptations to avoid a temporarily intolerable environment. Some close their shells and wait for conditions to return to normal. Others dig into the muddy bottom and wait. However, if the condition lasts too long they cannot survive. When possible, the nekton leave a poor environment. When they fail to leave, mass fish kills may occur.

Modern man is concerned about being able to produce enough food to nourish its expanding population. The question of how much food can ultimately be obtained from the sea is controversial. The ocean receives more than twice as much solar energy as land, and solar energy is the prime source of biological productivity. These two facts suggest that the ocean's productivity should greatly exceed that of the land. However, most of the sea is a biological desert. Higher productivities are found when runoff from the land or upwelling of nutrient-rich deep water fertilizes the euphotic zone, and stimulates the growth of marine plants. At the present high level of exploitation the fisheries of the world produce a small fraction of the human food requirements, and there is some risk that they may supply less in the future because of over-fishing.

Oyster farming has been practiced for hundreds of years. Oysters are well suited for farming because their spawn can be collected and used for "seeding" new areas of cultivation. The eggs of oysters develop into a nektonic larval form which attach themselves to a clean surface where they grow into adult oysters. After the oysters have attached themselves, it is easy for the farmer to move the clean surface, which he has provided, to another area. As in land farming, the farmer must protect his crop from predators.

Fish farming is more difficult because the fish are mobile and the species desired may not breed in the areas desirous for farming. The young fish may have to be caught at sea and transferred to restricted areas suitable for farming.

In both land and ocean types of farming the animals require food of some sort. The average annual yield is 520 tons per square mile for milk fish in Taiwan where the fish ponds are fertilized. In Indonesia, where sewage is put into the ponds in place of commercial fertilizer, the annual yield reaches 1,300 tons per square mile.[15]

4.4 CHEMICAL OCEANOGRAPHY

More than sixty different elements are present in sea water in measurable quantities. Many of the elements have been measured in marine organisms before they have been detected in sea water because the marine organisms contained higher concentrations of the elements than the sea water in which they live.

The ratio of nine dissolved, so called, "conservative" elements of sea water for an area is essentially constant although the total salinity may vary considerably.

Six major elements account for more than 90 percent of the salts in solution. These elements are: chlorine (19,000 ppm), sodium (10,500 ppm), magnesium (1,350 ppm), sulfate (885 ppm), calcium (400 ppm), and potassium (380 ppm). Three minor elements: bromine (65 ppm), strontium (8 ppm), and boron (4.6 ppm) complete the nine "conservative" constituents of sea water. If the concentration of one of the nine constituents is known, the others can be calculated.

Although carbon has an average concentration of 28 ppm, its ratio to the "conservative" elements is not constant. The other minor elements, (which are also not "conservative") are silica (3 ppm) and fluorine (1 ppm). Most of the remaining constituents of sea water; including dissolved gases, organic compounds, and particulate matter occur in differing ratios which are in part due to biological reactions.

Common trace elements in sea water are: nitrogen (0.5 ppm), lithium (0.17 ppm), rubidium (0.12 ppm), phosphorus (0.07 ppm), iodine (0.06 ppm), iron (0.01 ppm), zinc (0.01 ppm), and molybdenum (0.01 ppm). All the other known naturally occurring elements are possibly present in sea water in concentrations of less than 10 parts per billion. The concentration of the trace elements varies considerably with location, time, and biological activity.

It should also be noted that the concentrations of the "conservative" salts may be altered where the sea water is diluted by river water because the ratios between the major elements in river water are not the same as in sea water.

For all practical purposes the elements in sea water are present as parts of chemical compounds. Sodium and potassium compounds are very stable, whereas magnesium and silicon compounds are relatively unstable. The relative stability of these compounds affects the composition of the ocean because it affects the residence time of the elements in the ocean. Some elements appear to be concentrated in the ocean while others are quickly passing through the ocean system.

Most of the gases within the ocean come from the atmosphere, but some rare gases come from radioactive decay processes within or beneath the ocean. The gases with the largest concentrations in the ocean are nitrogen, oxygen, and carbon dioxide. Helium, neon, argon, krypton, and xenon are also present in minor amounts.

The concentration of a gas in sea water generally depends upon the temperature, type of gas, and salinity. Oxygen and carbon dioxide react with the marine environment, and therefore their concentrations vary independently of the above factors. As mentioned previously, the minimum oxygen zone is due to biological activity. Carbon dioxide is involved in biological activity, but also reacts directly with the somewhat alkaline sea water in the presence of certain cations such as calcium and magnesium. The following reversible reaction forms carbonates and bicarbonates:

$$\underset{\text{Carbon dioxide}}{CO_2} + \underset{\text{Water}}{H_2O} \rightleftharpoons \underset{\text{Bicarbonate}}{H_2CO_3}$$

$$\updownarrow$$

$$\underset{\text{Hydrogen ion}}{2H^+} + \underset{\text{Carbonate ion}}{CO_3^=} \rightleftharpoons \underset{\text{Bicarbonate ion}}{HCO_3^- + H^+}$$

As the demand for carbon dioxide by plants increases, more of this gas is released. When respiration releases carbon dioxide the reaction reverses and absorption of carbon dioxide occurs. The reaction also helps to regulate the hydrogen ion activity, and therefore the pH of the water which remains fairly constant. Sea water is a buffered solution which normally has a pH near 8. Higher or lower pH values may occur where the ocean circulation is restricted or the water is diluted.

Very little is known about the distribution of dissolved organic matter in sea water except for nutrients, and compounds of nitrogen and phosphorus. Other dissolved organic compounds found in sea water are organic carbon, carbohydrates, proteins, organic acids, amino acids, and vitamins. Concentrations of these are generally less than 6 mg per liter (ppm).

Some of the more important chemical reactions which control the composition of sea water take place at the boundaries with the atmosphere and the bottom sediments. These reactions are difficult to study and therefore are generally poorly understood.

One of the reactions not clearly understood is the formation of manganese deposits (fig. 4.11). These deposits are present in the Atlantic and Pacific Oceans. At some places they are in the form of nodules which apparently formed around a nucleus while resting at the sea bottom. At other places the deposits form a crust on the sediment bottom. So far the reason for the differences in form of deposit are unexplained as well as how the deposits choose their particular location. How a spherical object formed at the boundary between a liquid and a solid without being rolled over occasionally is not clear. It could be that the nodules were moved during formation by means not presently obvious.

Figure 4.11. Manganese nodules on a conveyor belt after discharge from nodule/water separator. (*photograph by B. J. Nixon, Deepsea Ventures, Inc.*)

4.5 ECONOMIC BENEFITS

Economic benefits from the ocean may come from the water itself, the shore, the bottom of the ocean, or beneath the bottom of the ocean. Man's knowledge of the ocean is limited such that one can only speculate concerning the economic benefits which may be derived from it.

The beaches and shorelines are valuable for recreational purposes. For example, large sums of money have been invested in cities along the coast of Florida because people like to live and vacation along this coastal area. Many of the cities along the coasts are ports which depend upon the ocean for transportation. Without the low cost transportation provided by the oceans the cost of goods in international trade would be much greater.

The oceans have a military value which is hard to estimate. In years past the ocean has provided a very effective barrier between nations, but this value has been somewhat lessened by our nuclear jet age. The ocean still provides a place to hide, support, and test new weapons.

As mentioned earlier there are many minerals in the ocean, but their concentrations are usually low. The ocean contains about 350 million cubic miles of water, and each cubic mile contains about 165 million tons of solids. However, to extract the solids in a cubic mile of sea water, 2.1 million gallons of water per minute must be processed for an entire year.

Sodium chloride (common salt) is extracted from the sea by using the sun to evaporate the water until the salt precipitates. In 1968 the value of this salt was about $10.00 per ton. Magnesium is extracted from sea water at a 1968 value near $665 per ton. Calcium chloride and bromine are extracted from sea water, and have monetary values between sodium chloride and magnesium.

In some areas the most valuable mineral which can be extracted from the ocean is fresh water. The methods of desalinization have steadily improved such that fresh water can be obtained from the sea at less than $1.00 per 1,000 gallons (figure 3.9). However, this price is still too high for irrigation purposes.

Commercial fishing in the ocean is important in order to feed the population of the world. It is of sufficient economic importance to some countries that they have extended their claims of exclusive fishing rights to 200 miles offshore.

Oil produced offshore amounts to about 18 percent of the world's oil production, and this percentage increases steadily.[16] Offshore crude oil production is expected to total over 25 million barrels per day by the year 1980, which may be one-third of the world output. Exploration drilling is occuring in sea water up to 460 meters (1,500 feet) deep, and it may occur in water 910 meters (3,000 feet) deep within the next five years. The value of the oil offshore is indicated by the money paid by one consortium for 3,427 acres off the coast of Louisiana, $94 million in 1968.[16]

Estimates by Lewis G. Weeks[16] indicate that offshore areas under less than 300 meters (1,000 feet) of water total 10.8 million square miles, 37 percent of which should be considered oil rich.[16] But only 17 percent of the 10.8 million square miles is classed as commercially attractive, and 2 percent is classed as top grade, bonanza-type.

The legality of who owns the fish and the minerals in the open ocean is presently being studied. Currently (1972) the ownership is ruled by the Geneva Convention of 1958 which holds that a nation has sovereignty on its continental shelf to a water depth of 200 meters (660 feet) or to where the depth of the superjacent waters admits of the exploitation of the natural resources.[16] This last clause may be interpreted several ways, and will probably create considerable conflict.

Until the ownership of a mobile fish is determined, aquaculture in the open sea will be impractical. Mining the deep ocean beyond the presently accepted continental limits is hazardous because ownership of land and minerals is not established. Deepsea Ventures Incorporated has made plans to mine manganese nodules 6,100 meters (20,000 feet) below the ocean surface.[17] The ownership of these nodules is not clear, except by the customary law of the high seas which includes the idea that what one can claim or control is yours. The problem of ownership becomes complicated when the company requires years to mine an area before a profit may be realized. Yet the oil companies have bought drilling rights from the United States to acreage well beyond the three mile Territorial Sea claimed by the United States, and beneath hundreds of meters of water with the full expectation of producing sufficient oil over a period of years to realize a respectable profit.

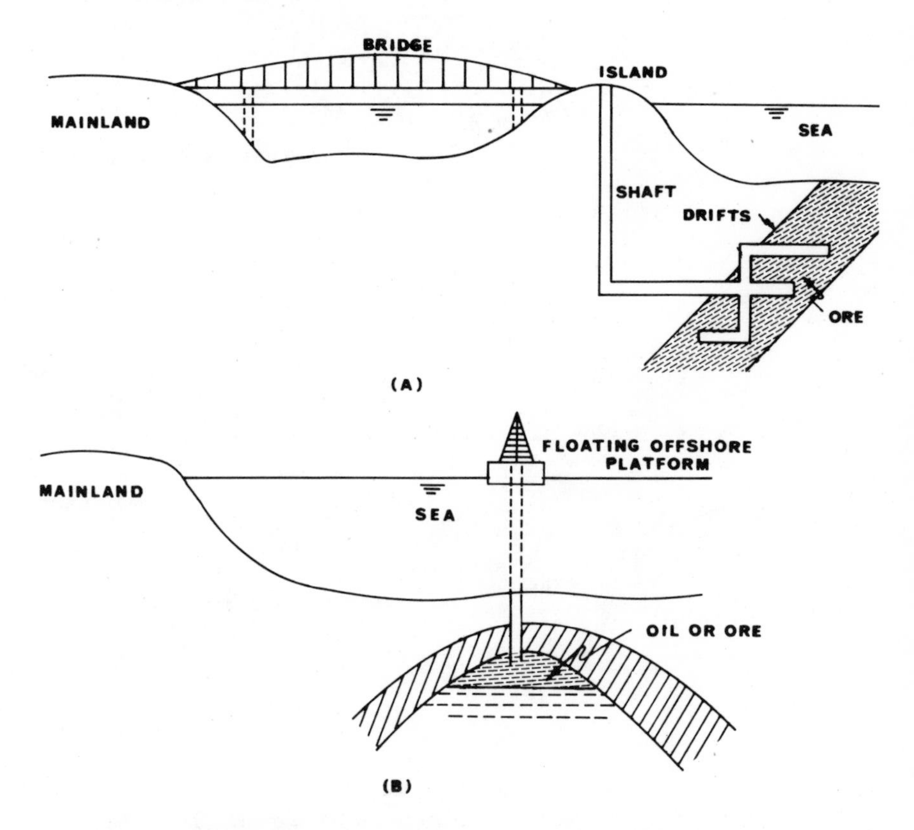

Figure 4.12. Extraction of minerals from beneath the ocean floor may take several forms, such as extending underground mining from the mainland via an artificial island (A) or supporting drilling equipment on a floating platform (B) above the mineral deposit.

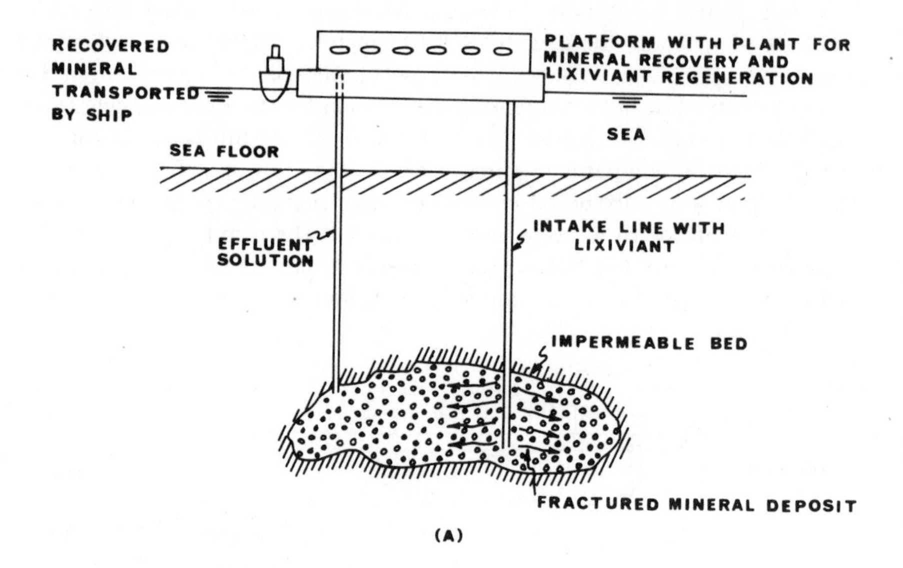

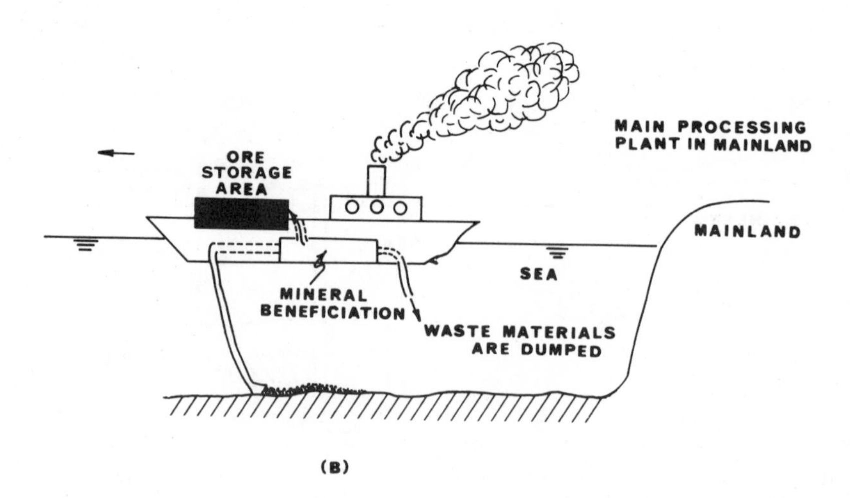

Figure 4.13. After entrance has been gained to the mineral body the drilling equipment may be replaced by production equipment (A) and solid minerals liquified for extraction. In cases where the minerals lie at the interface between the water and the sediment, the minerals may be raised to the ship by pumping or dredging (B). Aboard the ship or production platform the waste material can be separated from the ore and returned to the dredging site.

In spite of the problems of ownership, sea conditions which require special equipment, the high cost of operation, and the difficult living conditions, many minerals have been and are being mined offshore (figures 4.12 and 4.13). Most of the dredging operations operate in depths of less than 30 meters (100 feet). Diamonds, gold, heavy minerals, iron sands, sand and gravel, shells, and tin have been dredged from the bottom of the sea.[18] The greatest value is in ordinary sand and gravel because of the large quantity which is mined. Sand and gravel dredged offshore is more expensive than that obtained on land, except where land transportation to the use site increases the cost until it is above that of sand and gravel obtained offshore.

Offshore underground mining is generally conducted from openings onshore. The exceptions are sulfur which is mined by the Frasch process via holes drilled from platforms offshore, and some coal mined from shafts sunk from artificial islands. Most offshore coal is mined from shafts onshore. Iron and tin are also mined from shafts onshore.

4.6 DUMPING WASTE AT SEA

Pollution of the ocean environment occurs when the concentration of some constituent in the ocean becomes detectably different from its normal concentration, and this results in an objectionable change in the ocean's ecological balance. Eutrophication[19] can be defined as an enrichment of the nutrients in the water. Such an enrichment may lead to increased algal growth, changes in the quality of the water and deterioration of fisheries.

If the per capita waste production increases at the same growth rate (4 percent) predicted for production of consumer goods, the amount of material waste collected by municipal and private agencies should rise to eight pounds per day per person by 1980.[20] The amount of sewage sludge generated annually in the Baltimore-Washington region is expected to increase from 70,000 tons (dry weight) in 1968 to about 166,000 tons in 1980.[20] Sewage sludge barged to sea from the New York City area totaled 99,000 tons (dry weight) in 1960 and is expected to rise to 220,000 tons annually in 1980.[20]

The land requirements to dispose of waste from the metropolitan areas along the sea coast are increasing, and the availability of land is rapidly decreasing. In addition to these two facts it should be noted that the price of land suitable for waste disposal is increasing such that disposing of waste at sea is becoming the least expensive means of disposal. The combination of these factors is creating considerable pressure to permit greatly increased waste disposal in international waters.

As recently as 1970, no legislative enactment specifically "provides that any agency, department, or official of the United States is vested with authority to prevent pollution of the seas from discharge of solid waste beyond the outer boundaries of the Territorial Sea".[20] The United States' Territorial Sea extends three miles seaward of a baseline near the level of low tide. In practice the U.S. Army Corps of Engineers issues permits to barge waste to sea after consulting other interested agencies. Requirements for selecting and monitoring the dumping site vary from district to district.

Current marine disposal operations include several types of wastes and methods of disposal. Most of the wastes discharged in the ocean in 1968 were dredge spoils generated by harbor dredging operations of the Corps of Engineers. Industrial wastes consisting of refinery wastes, spent acids, pulp and paper mill wastes, chemical wastes, oil drilling wastes, and waste oil were second in quantity of those dumped at sea. Between 1946 and 1967 packaged radioactive wastes were dumped at sea by the U.S. Atomic Energy Commission, but this practice has now stopped. Other nations are continuing to dump radioactive wastes in international waters. New York City disposes of construction and demolition debris off the Atlantic Coast. Military explosives and chemical warfare agents are sunk in water depths greater than 1,830 meters (6,000 feet). The Dillingham report[20] located 281 areas designated for ocean disposal, which received approximately 62 million tons of wastes during 1968.

Little is known about the environmental effects of disposing solid and liquid wastes in the ocean. More is known about the immediate effects than the long-term effects because funds have been mostly available for short-term studies to answer specific questions.

Dredge spoils discharged at sea cause increased sedimentation and turbidity at the disposal site. The effect upon the biota of the area depends upon the type of organisms naturally present in the area, and their tolerance to environmental changes. The increased turbidity decreases light penetration, and thus the rate of photosynthesis. It also reduces the ability of animals to locate food visually. The sedimentation may destroy spawning areas, food supplies, and vegetation. Organic matter may be trapped in the sediments resulting in anaerobic bottom conditions. Often the dredge spoils are polluted, but the effect of this is not definitely known.

The reports of effects caused by industrial wastes discharged at sea indicate that these can usually be disposed of with little short-term observable effect on the biota of the area.[20] The long-term effects apparently have not been studied. But, as noted earlier, the concentration of toxins by organisms has been observed to affect higher forms of life in the food chain well outside the dumping site. The ability of a toxin to remain toxic over a long period of time should be studied in relation to its immediate effect on the biota.

Many highly toxic substances are dumped into the ocean, but they are dumped in such a manner that they are quickly diluted to the extent that the biota are not significantly affected. In some cases the lack of apparent effect may be due to the absence of visible life in the area.

The buffering action of ocean water rapidly reduces the acidity of acid-iron disposal operations. It was reported that the concentration of bluefish increased around the outer boundaries of one area of acid-iron disposal.[20]

Liquid wastes which are denser than sea water have not been observed to sink in a plume to the bottom where their concentrations might remain high enough to be toxic. These wastes sometimes sink over an area and become diluted to the density of a lower layer of water where they may be distributed over large areas by currents and eddy diffusion.

The selection of an ocean dumping site for a particular type of waste must be studied individually. Water circulation, rate of sedimentation, bottom characteristics, and biota must be studied in order to predict the effect the waste will have on a selected area. The deepest location farthest from shore is not necessarily the best site for disposal of a particular waste material.

Certain waste materials suitable for constructing artificial reefs may serve to improve the fishing, provide interesting areas for SCUBA diving, and serve as breakwaters. Rubber tires appear to make the most successful long-lived reef for the attachment of benthic life which attract fish. Rubber tires anchored near the surface are effective at reducing wave action. Automobiles make good reefs, but they are expensive to put in place, and in a few years the chemical action of the sea water destroys the cars. Bales of refuse may make good habitats for fish, and may provide food as well.

Recycling waste material (see Chapter 6) rather than dumping it at sea would be more beneficial to mankind in some situations. Such a practice may become necessary and economical in the future.

4.7 SUMMARY

The oceans are very much a part of our environment. The continents cover only about 29 percent of the earth's surface, while the oceans cover the rest. Man, with few exceptions, lives at the interface between the atmosphere and the land. For the most part he is dependent upon this interface for his food and water. The climate of this interface is largely controlled by the interaction of the sun and water at another interface, namely, that

Figure 4.14. Chamber of the underwater habitat. (*photo by Dave Woodward, International Underwater Explorers Society*)

between the ocean and the atmosphere. A large scale change in the ocean surface has a large scale effect on the climate which may have good or bad consequences for man.

Because the ocean is fluid and circulates water from all depths to the surface, the entire body of water must be considered as potentially at the surface. Further, the chemical and biological activity of the ocean may change substances put in it. Over long or short periods of time innocuous substances may become very objectionable. Stopping or reversing some reactions also takes time during which the effects of the changes must be endured.

Consideration of the above remarks should clearly indicate that man should be careful as to how he uses the ocean, and whenever possible should know the effect of his actions before he takes them. Of course, there is no reason why man should not exploit the ocean for the benefit of mankind, but the individual or group must consider the global long-term effects as well as their immediate desires.

Man has used and will continue to use the ocean and the earth beneath it as a source for petroleum, solid minerals, chemicals, food, and water. He will continue to use the ocean for transportation and waste disposal. Perhaps in the future he will use it as a place to live (figure 4.14). Consumption of resources economically obtainable on land will eventually reduce these reserves to the extent that obtaining resources offshore will become necessarily economical.

Increasing our knowledge of the ocean, and improving the equipment required to obtain knowledge and minerals from the ocean contributes to the future when more complete utilization of the ocean resources will be necessary. Many years will be required to make the technical advancements necessary to economically obtain ocean resources without disrupting the environment.

4.8 REFERENCES

1. Raymont, J. E. G. (1963) "Plankton and Productivity in the Oceans", Pergamon Press, New York, N.Y.
2. Isaacs, J. D. (1969) *The Nature of Oceanic Life*, Scientific American, volume 221, pp. 147–162.
3. von Arx, W. S. (1962) "An Introduction to Physical Oceanography", Addison-Wesley Publishing Co., Inc., Reading, Mass.
4. Forch, C., Knudsen, M., and Scrensen, S. P. L. (1902) "K.danske vidensk. Selsk.", volume 6.
5. Jacobsen, J. P. and Knudsen, M. (1940) Assoc. d'Oceanog. Phys., U.G.G.I. Publ. Sci., No. 7.
6. Knudsen, M. (1901) "Hydrographical Tables" 2nd ed. 1931, Copenhagen: G.E.C.G.A.D., London: Williams and Northgate.
7. Ross, D. A. (1970) "Introduction to Oceanography", Appleton-Century-Crofts, New York, N. Y.
8. McLellan, H. J. (1965) "Elements of Physical Oceanography", Pergamon Press, New York, N.Y.
9. Neumann, G. and Pierson, Jr., W. J. (1966) "Principles of Physical Oceanography", Prentice-Hall, Inc., Englewood Cliffs, N. J.
10. Pierson, Jr., W. J., Neumann, G., James, R. W., (1955) "Practical Methods for Observing and Forecasting Ocean Waves by Means of Wave Spectra and Statistics", U.S. Navy Hydrographic Office Publication 603.
11. McHugh, J. L. (1967) *Estuarine Nekton*, in "Estuaries" edited by Lauff, G. H., Amer. Assoc. Adv. Sci., Washington, D.C., Publication No. 83, pp. 581–620.
12. Horn, R. A., (1969) "Marine Chemistry", John Wiley & Sons, New York, N. Y.
13. Sverdrup, H. V. (1955) *The Place of Physical Oceanography in Oceanographic Research*, Journal of Marine Research, volume 14, pp. 287–294.

14. Jones, E. P. (1971) *DDT Stopped, Suit Dropped*, Science, volume 173, p. 38.
15. Pinchot, G. B. (1970) *Marine Farming*, Scientific American, volume 223, pp. 15–21.
16. Gardner, F. J. (1971) *Oil's Offshore Activity Suppressed by Economics, Political Decisions*, Oil and Gas Journal, volume 69, pp. 97–103.
17. Kaufman, R., and Rothstein, A. J. (1970) *Recent Developments in Deep Ocean Mining*, Marine Technology Society, 6th Annual Preprint, volume 2.
18. Cruickshank, M. J. (1969) *Mining and Mineral Recovery*, Undersea Technology Handbook/Directory 1969, Compass Publications, Inc., Arlington, Va.
19. Likens, G. E., Bartsch, A. F., Lauff, G. H., and Hobbie, J. E. (1971) *Nutrients and Eutrophication*, Science, volume 172, pp. 873–874.
20. Smith, D. D. and Brown, R. P. (1971) "Ocean Disposal of Barge-Delivered Liquid and Solid Wastes from U.S. Coastal Cities", U.S. Environmental Protection Agency, Publication No. SW-19c.

Chapter 5

MINERAL RESOURCES

5.1 INTRODUCTION

Having looked at water as one of the most valuable resources in the previous chapters, it is appropriate that we now consider mineral resources which are equally valuable and essential for maintaining our huge population and for operating our civilized society. Not too many people realize the extent to which our modern industrial age and civilization have become dependent upon minerals. One only has to look around in order to appreciate the significant contribution that mineral resources have made to our existence by providing the necessities of life, health, security, economy, and well-being.

Man, by his ability to utilize the mineral resources of the earth, has made a tremendous progress in a relatively short geologic time. At the dawn of civilization, he used crude stones, flint, chert, and bone implements which were the only useful raw materials available. Chaldeans, Babylonians, and Egyptians made extensive use of clay for pottery, bricks, and tiles. Building stones were widely used as exemplified by the Pyramid of Gizeh containing 2,300,000 blocks of stone each weighing 2½ tons. Neolithic man was acquainted with native metals such as gold, copper, and tin, mostly from placer sources. Herodotus (484?–425 B.C.), Theophratus (372–287 B.C.), Strabo (A.D. 19), Avicenna (980–1037), and Agricola (1494–1555) in his famous De Re Metallica,[1] have all described the occurrences, methods of extraction, and uses of metals, gemstones, earths, and salts. At the termination of the Dark Ages, the major metals in use were iron, copper, lead, tin, gold, silver, and mercury.

With the advent of the Industrial Revolution three centuries ago, man learned to produce iron from ores using coal and mechanical power. Since that time the use of metals increased rapidly and each progressive step in the history of man brought about increasing demand for old metals and the discovery and development of newer ones. During World Wars I and II, more metals and minerals were used than in all preceding history.

Today more than one hundred minerals are involved in international trade, some being necessary and vital for our essential industries, while others contribute to unessential industries or luxuries. Regardless of the end use, it is evident, that with increasing population and with the industrialization of developing countries, the demand for our mineral resources will continue to increase.

However, man in his persistent efforts to satisfy his insatiable demand for mineral resources, has mostly exploited them without regard to consequence (see section 1.1). It would be fruitless to deny that the mineral industry by its very nature contributes to certain pollution problems. Open pit mining, underground operations, hydraulic mining, oil and gas production, metal smelting, fuel refining and all other extractive efforts, disturb to some degree our environment. Some damage to the environment is inevitable even with the best extraction procedure and land restoration methods. The mined ores are moved and handled several times before they are processed, thus creating noise, dust, and other forms of air pollution. Mining wastes and drainage usually contribute to stream pollution and create a solid waste disposal problem. Smelting and refining of ores and concentrates invariably add to air pollution. Seepages of oil and brine during oil and gas production also constitute special pollution problems and so do mine fires, surface subsidence, drainage from underground mines, and mining on the continental shelf.

It is for these reasons that the mineral industry is identified by a large segment of the uninformed public as a major contributor to the overall pollution problem. This is far from the truth. In the first place, only a small percentage (about 0.14 percent) of the total land in the United States has been adversely affected by mining. Moreover, practically no public attention has been drawn to the extensive as well as expensive efforts made by the mineral industry to minimize the environmental impact of its operations. Many extractive companies take great pride in the responsible manner in which they have conducted their business. Their efforts in the environmental control field have gained national recognition and praise from government agencies as well as the public.

It is not the intention of the authors to cover in detail the entire field of mineral resources and the associated environmental problems. For such detailed coverage, the reader can refer to any of several excellent books available on the subject.[2–8] Rather, the primary objectives are to develop the importance of the resources in our minerals-based civilization, to point out the environmental problems associated with the extraction of these badly needed resources, and to bring out the important concept of extracting mineral wealth without creating environmental poverty through the use of technology and a healthy public attitude.

5.2 EARTH AS A SOURCE

Our earth provides us with water, food, energy, and minerals, all of which constitute our *natural resources*. We need these resources to sustain life and our multifarious civilization. Of these, some such as food, forests and fibers, once planted and cultivated, can be regenerated seasonably and we refer to these as *renewable resources*. On the other hand, natural resources like minerals, gas and oil are called *nonrenewable resources* because the deposits containing them have limited supplies which, after long enough exploitation, are totally exhausted.

The above concepts bring up a very crucial question about our civilization. Does the earth contain sufficient quantities of our depleting mineral resources to support and expand our complex civilization? The answer to this question is not simple since estimating the capabilities of the earth to supply mineral resources involves a large number of uncertainties. In the first place, numerous estimates of the available reserves and locations of different mineral deposits are not reliable. Secondly, whether a particular type and grade of mineral material constitutes or can become an ore (a mineral deposit that can be exploited at a profit), depends on numerous economic factors that vary in time. With a breakthrough in mining or extractive technology or an increase in price, worthless rock can be converted to an ore overnight.

At this point we are confronted with another crucial question. Are there any limits to the grades of ore that can be processed? From a theoretical as well as a technological viewpoint there is no lower limit as was illustrated beautifully by the isolation of plutonium from uranium metal at Oak Ridge during World War II. In this case, the content of plutonium in uranium metal amounted to only a few parts per million and the extraction involved a very complex chemical processing scheme which obviously was very expensive and uneconomical.

Today, we are economically extracting copper from ores containing around 0.6 percent copper, and there is no doubt that if there is a critical need for the metal in the distant future, we will continue to extract copper from lower and lower grade ores containing from 0.1 percent to 0.01 percent copper. However, such an achievement could be possible only if a suitable metallurgical process and the required energy for processing are available. Keeping this in mind any material in the earth's crust can be considered a potential source if sufficient energy is readily available. Thus from a technical standpoint, we can safely state that our resource potential will be proportional to the available energy reserves. Since all the technical reports concerning the availability of energy to man indicate an enormous reserve to satisfy the highly industrialized world for many years, it is evident that we will continue to obtain the required metals from lower and lower grade ores and ultimately from mediocre metal-containing-rocks such as shale and granite.

Nonrenewable resources are considered as one of the most important assets of a nation. The future of our civilized society depends directly on a continuing supply of these resources. According to Lovering[9] the total volume of workable mineral deposits is only a very small fraction of one percent of the earth's crust. In addition, every mineral deposit can be considered to be the result of the existence of given geologic conditions at a time in the past. Since mineral deposits have a fixed position they must be mined where they occur. Also, once the deposit which has limited quantity of ore is mined out, no new ore will grow in its place and thus it is a depleting asset.

It is true that some of the more durable metals like copper, iron, and lead are recovered as scrap, but their recovery is only partial and they are permanently consumed after several reclamation cycles. Water is one of the few nonrenewable resources which is not permanently consumed and thus becomes a *reusable resource.* However, as we deteriorate more and more the quality of water, it becomes more and more expensive to reuse it (see Chapter 3).

It is also clear that rich and easily treated mineral resources constitute valuable but short-lived assets of a nation; rise and fall of empires having been attributed to the discovery and depletion of these nonrenewable resources. The current growth taking place in Australia, Alaska, and Indonesia are fine examples illustrating the validity of the above observation. It is also understandable that the mining and operating costs increase with increasing depths and decreasing grades. However, with technological improvements and the economic advantage gained through large scale operation, it is possible to offset increasing costs of operating lower grade deposits and realize an economically feasible operation. Finally, the reader should be aware of the fact that our complex civilization and our expanding industries place insatiable demands on our limited mineral resources and in some cases we have found suitable substitutes for essential commodities. But, desired substitutes for a large number of basic metals and non metals are difficult to find and we have to rely on lower and lower grade ores and improved technology to keep on producing the commodities required by our increasing population and expanding industries.

In estimating the availability of mineral resources, geologists have designated several types of reserves. A *measured reserve* indicates a mineral deposit for which all the pertinent information as to the tonnage and grade is available. However, if the measurements are not exact and involve some guess work the resource becomes an *indicated reserve.* On the other hand, if the estimate is based on geological reasoning and experience and does not involve any measurements then it is known as an *inferred reserve.* It should be noted that these reserve terms refer only to economically workable deposits and not to identified deposits which may become economically workable in the future. Such deposits are called *potential resources.*

On the basis of the above reasoning, it would not be wise for us to speculate with known and unknown reserves. Rather, it would be meaningful for us to examine some important characteristics of mineral resources; to look at current and future demands for various metals, minerals, and energy sources; and to suggest ways and means of improving the future outlook for our mineral-based economy.

In our discussion we shall consider mineral resources to include almost anything of economic value that can be extracted from the earth such as metals, industrial minerals, fossil fuels, water, and natural gases. Table 5.1 shows the types of mineral resources as classified by Skinner[10] on the basis of their use. The first group consists of *metallic mineral resources* which are chemical elements useful to man because of their metallic properties. These have been further subdivided on the basis of their relative abundance in the earth's crust. Thus, *abundant metals* include iron, aluminum, chromium, and titanium while rarer metals such as copper, lead, zinc, gold, and molybdenum are grouped under *scarce metals.*

TABLE 5.1.

Types of mineral resources*

Metallic Mineral Resources	Nonmetallic Mineral Resources
Abundant metals	*Minerals for chemical, fertilizer, and special uses*
Iron, aluminum, chromium, manganese, titanium, and magnesium	Sodium chloride, phosphates, nitrates, and sulfur, etc.
Scarce metals	*Building materials*
Copper, lead, zinc, tin, tungsten, gold, silver, platinum, uranium, mercury, molybdenum, etc.	Cement, sand, gravel, gypsum, asbestos, etc.
	Fossil fuels
	Coal, petroleum, natural gas, and oil shale
	Water
	Lakes, rivers, ground waters and oceans

* After Brian J. Skinner, Earth Resources (*C*) 1969, p. 10. Reprinted by permission of Prentice-Hall, Inc., Englewood Cliffs, N.J.

TABLE 5.2.

Average composition of the earth's crust according to Clarke and Washington[13]

	Percent		Percent
Oxygen	46.710	Tungsten	0.005
Silicon	27.720	Lithium	0.004
Aluminum	8.130	Zinc	0.004
Iron	5.010	Columbium and tantalum	0.003
Calcium	3.630		
Sodium	2.850	Hafnium	0.003
Potassium	2.600	Lead	0.002
Magnesium	2.090	Cobalt	0.001
Titanium	0.630	Boron	0.001
Manganese	0.100	Beryllium	0.001
Barium	0.050	Molybdenum	0.0001
Chromium	0.037	Arsenic	0.0001
Carbon	0.032	Tin	0.0001
Zirconium	0.026	Cadmium	0.00001
Nickel	0.020	Mercury	0.00001
Vanadium	0.017	Silver	0.000001
Cerium and yttrium	0.015	Selenium	0.000001
Copper	0.010	Gold	0.0000001

The second group consists of *nonmetallic mineral resources* which are divided into four subgroups on the basis of their use. These include (a) minerals useful in chemical and fertilizer industries, (b) building materials, (c) fossil fuels, and (d) water.

It should be noted that some of the above mentioned mineral resources are found in nature as native metals or elements (diamonds, sulfur, copper, gold, mercury, and helium); while others like quartz (SiO_2), calcite ($CaCO_3$), kaolinite ($H_4Al_2Si_2O_9$), oil, and natural gas, are compound minerals whose compositions can be expressed by chemical formulas. However, the fundamental distinction among the minerals rests on their varying internal structures and not on their chemical composition as illustrated by diamond and graphite which are both composed of pure carbon but occur in different crystal forms. Diamond is the hardest substance known while graphite is soft and greasy; diamond is used as an abrasive while graphite acts as a lubricant.

5.3 ABUNDANCE AND AVAILABILITY OF MINERAL RESOURCES

Since the availability of a metal for commerce or industry depends to a large degree on its prevalence in the earth's crust, a knowledge of the relative abundance of the elements in the crust of the earth should be of interest to us. Classical studies on the abundance of elements by Mason,[11] Goldschmidt,[12] and Clarke and Washington,[13] have revealed that the earth is composed of 88 different elements even though 103 elements are now known, the elements above uranium (number 92 in the periodic table) being man-made. Table 5.2 gives the average chemical composition of the earth's crust in terms of elements as estimated by Clarke and Washington on the basis of about 5000 chemical analysis.

As indicated, the first eight elements—oxygen, silicon, aluminum, iron, calcium, sodium, potassium, and magnesium—account for 99 percent of the earth's crust while eleven elements account for 99.46 percent. The remaining elements including valuable metals such as antimony, chromium, copper, gold, lead, mercury, nickel, silver, tin, tungsten, and zinc, together

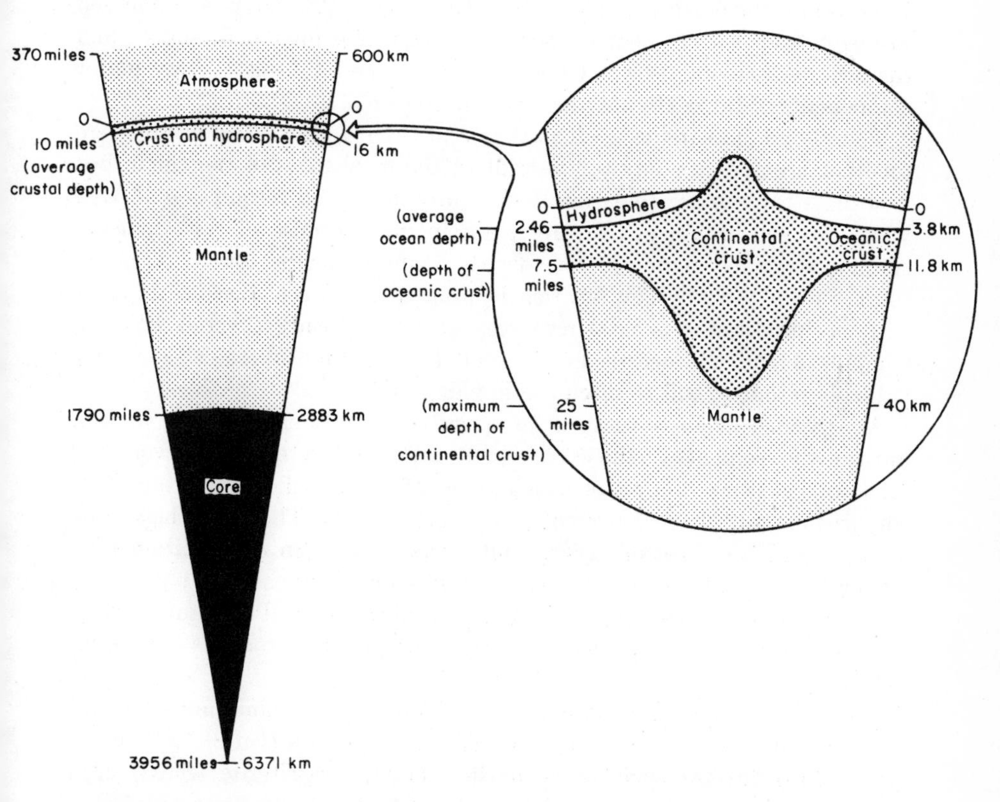

Figure 5.1. Concentric structure of the earth zones. (After B. J. Skinner, Earth Resources, (C) 1969, p. 17. *Reprinted by permission of Prentice-Hall, Inc., Englewood Cliffs, N.J.*).

constitute less than 0.5 percent. It is thus quite clear that geological processes of concentration are essential for collecting these diffuse elements into economic and workable mineral deposits. It is significant to note that zirconium is more abundant than copper or zinc while unfamiliar metals such as hafnium, columbium, and tantalum occur in greater amounts than lead and tin which are well-down on the list. As far as mercury, silver, and gold are concerned, they are present only as traces in the earth's crust but it is fortunate indeed that nature has concentrated them into workable deposits.

It is estimated that the earth has a mass of 6.5×10^{21} tons of which the earth's outer crust (10 miles deep) accounts for 0.375 percent of the mass. The earth also consists of a *mantle* of dense oxide minerals of iron and manganese as well as a metallic *core* containing principally iron and nickel, these two together accounting for more than 99.6 percent of the total mass. In addition, the earth is made up of a *hydrosphere* (the body of condensed water on and near the earth's surface) accounting for 0.025 percent and the *atmosphere* amounting to 0.0001 percent of the total mass. These major components of the earth are illustrated in Figure 5.1.

The above data indicates that the crust, hydrosphere, and atmosphere together account for only 0.4 percent of the total mass of the earth, but it is these three zones that supply our current mineral resources and from which we will obtain our future requirements. Because of their inaccessibility, we cannot seriously consider the mantle and the core as potential sources even though they are rich in mineral resources. Because of their extreme importance as vital resources we have already covered water in Chapter 3 and the oceans in Chapter 4. The atmosphere, also a very valuable resource, will be discussed in Chapter 7.

As stated previously, with a few possible exceptions of native metals, most of the elements in the earth's crust are found in minerals. Over 1600 mineral species are known of which about fifty are rock-forming minerals, comprising more than 95 percent of the earth's crust. These are the *silicate minerals* which consist of oxygen and silicon usually in combination with one or more metallic elements such as aluminum, iron, sodium, potassium, calcium, and magnesium often forming complex chemical structures. Most of these silicates constitute gangue, the term applied to the waste material in ores.

About two hundred minerals are considered as economic minerals and these include, sulfides such as galena (PbS), chalcocite (Cu_2S), and sphalerite (ZnS); *oxides* such as hematite (Fe_2O_3), cassiterite (SnO_2), and cuprite (Cu_2O); *carbonates* such as calcite ($CaCO_3$), dolomite ($MgCO_3$), and strontianite ($SrCO_3$); *sulfates* such as barite ($BaSO_4$) and celestite ($SrSO_4$); and *halides* such as halite ($NaCl$) and fluorspar (CaF_2).

Is it possible for technology to always guarantee a supply of needed mineral resources? Some geologists like Cloud[15] believe that these resources are limited and in spite of developing technology it will not be possible for us to overcome this deficiency. Alternatively, according to many economists like Barnett and Morse,[16] and scientists like Weinberg,[17] these essential resources are unlimited and available to man for a long time to come. This difference of opinion may be attributed to the definition of resources. On one hand, Cloud's interpretation of mineral resources is quite rigid and states that the amount of any metal contained in the earth's crust is limited and when this is used up, there will be no more left. He also feels that recycling of the metal, though possible, is expensive, inefficient and the metal will eventually be permanently lost. On the other hand, economists and technological optimists like Weinberg take a liberal definition of resources and believe that if man runs out of one metal, he will use substitutes, such as aluminum instead of copper. They also feel that ultimately, man would be able to sustain a complex and efficient civilization by using only the most abundant elements such as hydrogen, oxygen, carbon, nitrogen, sodium, potassium, iron, aluminum, silicon, phosphorus, and titanium.

This latter thesis assumes that unlimited energy will be available to extract the mineral resources. It is obvious that lower grades of resources will require greater amounts of energy for extraction and energy will be the key to the future. Since the world supply of fossil fuels is limited, we will have to rely on other sources of energy such as solar and nuclear. Of these two, the solar source will be very difficult to harness and it is apparent that we would have to rely on nuclear energy, either of the fission or fusion type, to provide us with inexhaustible energy.

A third outlook on the future of metal resources is projected by Wright[18] who believes that the problem is not whether there are adequate reserves or not. The important consideration is the growth rate of consumption in the light of increasing use of plastics and other materials for certain uses which makes the demand for metals tolerable. Under these conditions, metals will be used less and less in those applications where their special properties are not essential and will become higher priced supermaterials or "noble materials" of the future just as precious metals are regarded today.

It must also be noted that the extraction of resources from lower and lower grade will be accompanied by increasing pollution which is essentially a by-product of affluence in addition to being enhanced by overpopulation.

In order to truly appreciate the extent and the economic importance of the mineral resources used by man, let us cast a look at the magnitude of

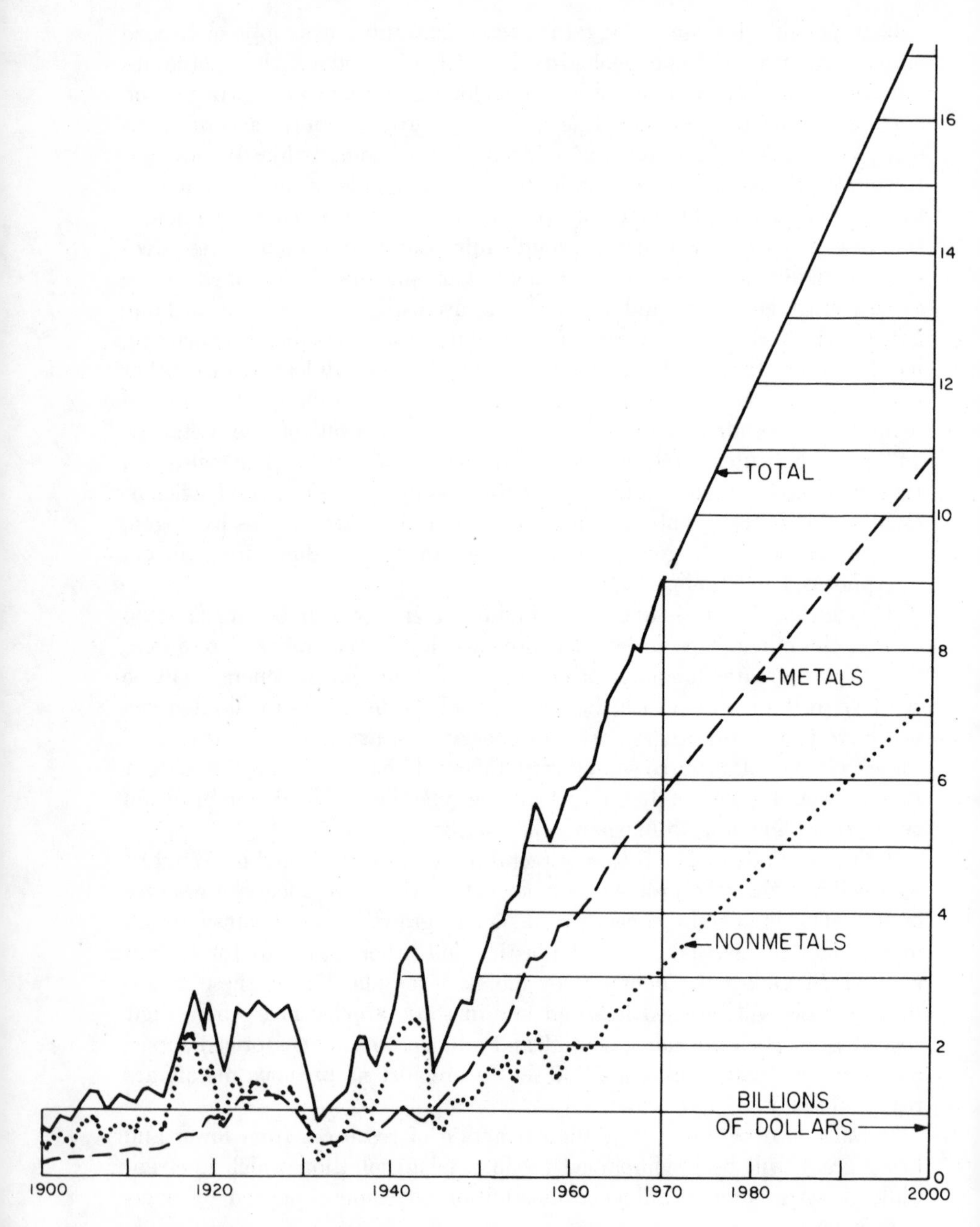

Figure 5.2. Total value of U.S. metallic and nonmetallic mineral production based on United States Bureau of Mines Figures and Projections to year 2000 by Engineering and Mining Journal.[23]

the minerals business in the United States. According to the information presented in Figure 5.2 it can be seen that the total value of the metallic and nonmetallic mineral production in the United States during 1970 amounted to about 9 billion dollars, the metals representing about 60 percent of the total value. If we now add the total value of all the energy resources (oil, gas, coal, and uranium) used in 1970 to the above figure, the total value of all mineral resources amounts to a staggering 28 billion dollars, undoubtedly a tremendous business contributing significantly to the overall economic health of the country. The projected volume of business in the year 2000, based on corresponding increases in population at a rate of doubling population every 37 years, is estimated at 23 billion dollars for metals and nonmetallic minerals and at about 60 billion dollars for the total value of all mineral resources including energy.

The reader is reminded that the above figures are for the United States alone and do not include the value of metals, nonmetallic minerals, and energy used by other free nations of the world and the Communist block countries. If we include all these figures then the total value may reach at least 100 billion dollars for 1970 and several hundred billion dollars for the year 2000. In the future, it will be the awakening of people in developing nations and unfolding of their industrial development that will multiply world requirements for metals and minerals.

From the above discussion, it is evident that the industrialization of the

TABLE 5.3.

Projected world demand for selected commodities equated to 1970 United States standard of living*

(Million short tons)

Metal or Commodity	U.S. Consumption in 1970	World Consumption in 1970	Projected World Consumption to Equal 1970 U.S. Standard
Iron (Metal)	98.00	500.0	1670
Copper	3.00	9.2	51
Lead	1.60	3.8	27
Zinc	1.80	6.2	30
Tin	0.09	0.2	2
Aluminum	5.10	12.0	90
Sulfur	9.30	20.0	160
Phosphorus	3.70	9.0	63
Potassium	3.80	11.0	65

* Projected on data from U.S. Bureau of Mines, 1970.

developing countries will be greatly affected by the availability of critical mineral resources. This fact can be readily appreciated if one examines Table 5.3 which shows the demand for these resources if all of the 3.7 billion people of the world of 1970 want to emulate the American standard of living.

At the current rate of annual world production, the figures mean the extraction of about four times as much iron, seven times as much copper, nine times as much lead, six times as much zinc, and ten times as much tin. To provide similar conditions in the year 2000 with its projected population, it would be necessary to double all the above figures. These statistics along with the problem of supplying the required raw materials would stagger anyone's imagination and it is obvious that our environment may have a very difficult time in supporting total industrialization of the world.

In looking at the future, it can be said that until the year 2000, the developed countries will probably be able to obtain the required mineral resources, either from domestic or foreign sources, since most of the underdeveloped countries will undergo only partial industrialization at best. However, by 2050, the world in general will face great difficulties in obtaining the needed natural resources. Hopefully, by then, man will have learned to work with substitute materials and make wise use of whatever mineral resources are still available.

Metallic Minerals

In comprehending the current metals and minerals requirements of the world and in projecting them for future years, we will choose the United States as an example since up-to-date facts and production figures are available. Also, such an analysis is useful because many developing nations of the world aspire to achieve living standards comparable to those of the United States. Under these circumstances, these data would provide meaningful guidelines for realistic projections of mineral production and consumption. The first realistic analysis of future mineral resources and requirements was made in 1952 by the President's Materials Policy Commission, later named the Paley Commission.[20] In the Paley Report, the United States mineral resources position until the year 1975 was appraised and this study, though not perfect, gave very valuable information on the state of our mineral resources.

The next important study was undertaken by Resources for the Future Inc., a non-profit group, who critically evaluated the future mineral requirements, demand for key metals and minerals, and adequacy of the resource base. The results of this classical appraisal were published in

1963 in "Resources in America's Future"[21] and is considered as one of the best and most complete appraisals available to date. Of equal importance are the publications of the U.S. Bureau of Mines: Minerals Yearbook,[22] and annual review and Minerals Facts and Problems,[23] published every five years.

In projecting the demand for mineral resources ten to thirty years hence, one should relate mineral requirements to population trends; consumption of basic mineral raw materials such as iron, copper, lead, zinc, sulfur, and fertilizer minerals such as phosphate and potash; and the current and future energy needs. The inclusion of details concerning all the mineral resources in beyond the scope of this chapter. However, a review of a few major metals and raw materials indicating the demand trends will suffice to give the reader a comprehensive picture of the availability of mineral resources in the future and the challenge that would be provided to the mining industry in order to meet the demands.

Figure 5.3 shows the United States requirements for the period 1960–20000 for important minerals and metals such as iron (a), aluminum (b), copper (c), and zinc (d). These graphs were adapted by the Engineering and Mining Journal[19] from a publication of Resources for the Future, Inc.[21] The *reserves* shown in the graphs have been estimated by the U.S. Bureau of Mines, U.S. Geological Survey or other sources prior to 1961. The *identified resources* represent important known reserves but which are too low in grade to be extracted at a profit using currently available technology, but containing sufficient metal values to be considered as important sources in the future. It is this type of resource which will constitute a challenge to the minerals engineer and the mining industry. The impetus in research activities during the last ten years has already narrowed the gap between reserves and the identified resources.

Non-Metallic and Fertilizer Minerals

A similar projection of the current and future demand for non-metallic and fertilizer minerals such as sulfur, phosphate and potash is illustrated in Figures 5.4 and 5.5. These figures were presented by Proctor and Beveridge[24] and are based on estimates by the U.S. Bureau of Mines[22] and Resources for the Future Inc.[21] The overall picture for sulfur reveals that the annual growth rate from 1960 to 1970 has been in excess of 4 percent and it is evident that if this growth rate continues, the future requirements for sulfur will be monumental. In this connection, it is gratifying to note that the current and future environmental considerations concerning the restriction of sulfur dioxide emissions from the smelting of base metals

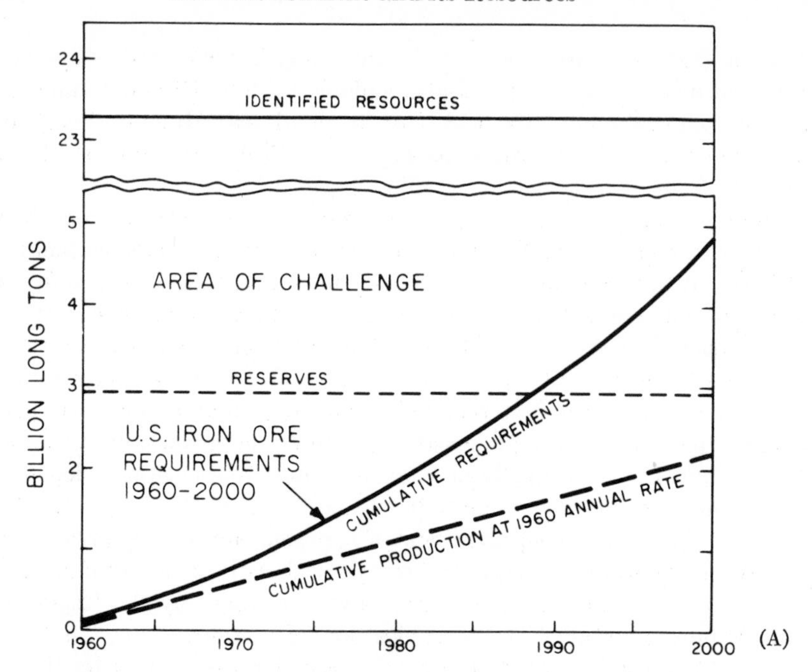
24
23
5
4
3
2
0
1960
1970
1980
1990
2000
BILLION LONG TONS
IDENTIFIED RESOURCES
AREA OF CHALLENGE
RESERVES
U.S. IRON ORE
REQUIREMENTS
1960-2000
CUMULATIVE REQUIREMENTS
CUMULATIVE PRODUCTION AT 1960 ANNUAL RATE

(A)

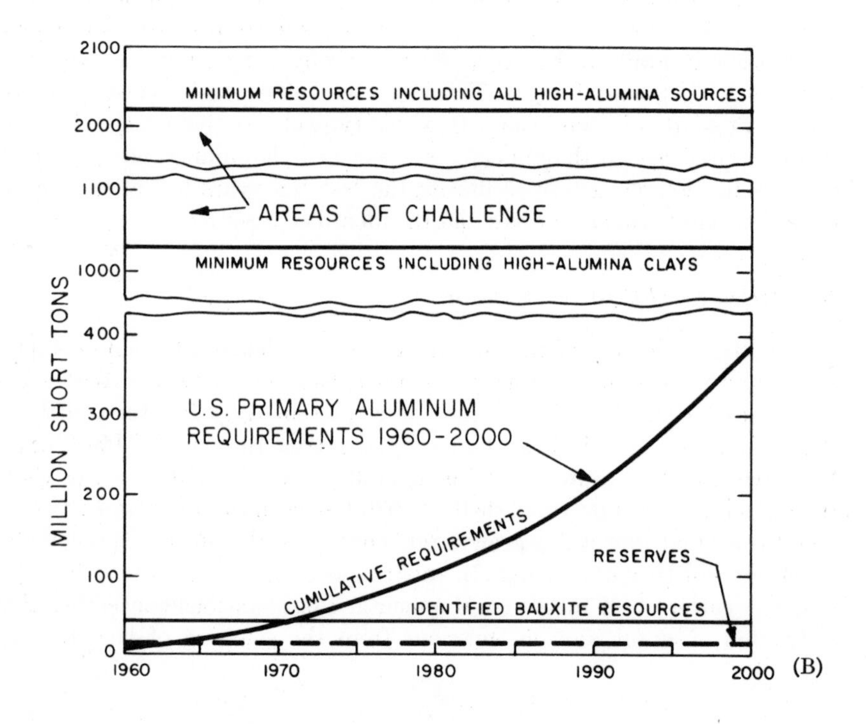
2100
2000
1100
1000
400
300
200
100
0
1960
1970
1980
1990
2000
MILLION SHORT TONS
MINIMUM RESOURCES INCLUDING ALL HIGH-ALUMINA SOURCES
AREAS OF CHALLENGE
MINIMUM RESOURCES INCLUDING HIGH-ALUMINA CLAYS
U.S. PRIMARY ALUMINUM
REQUIREMENTS 1960-2000
CUMULATIVE REQUIREMENTS
RESERVES
IDENTIFIED BAUXITE RESOURCES

(B)

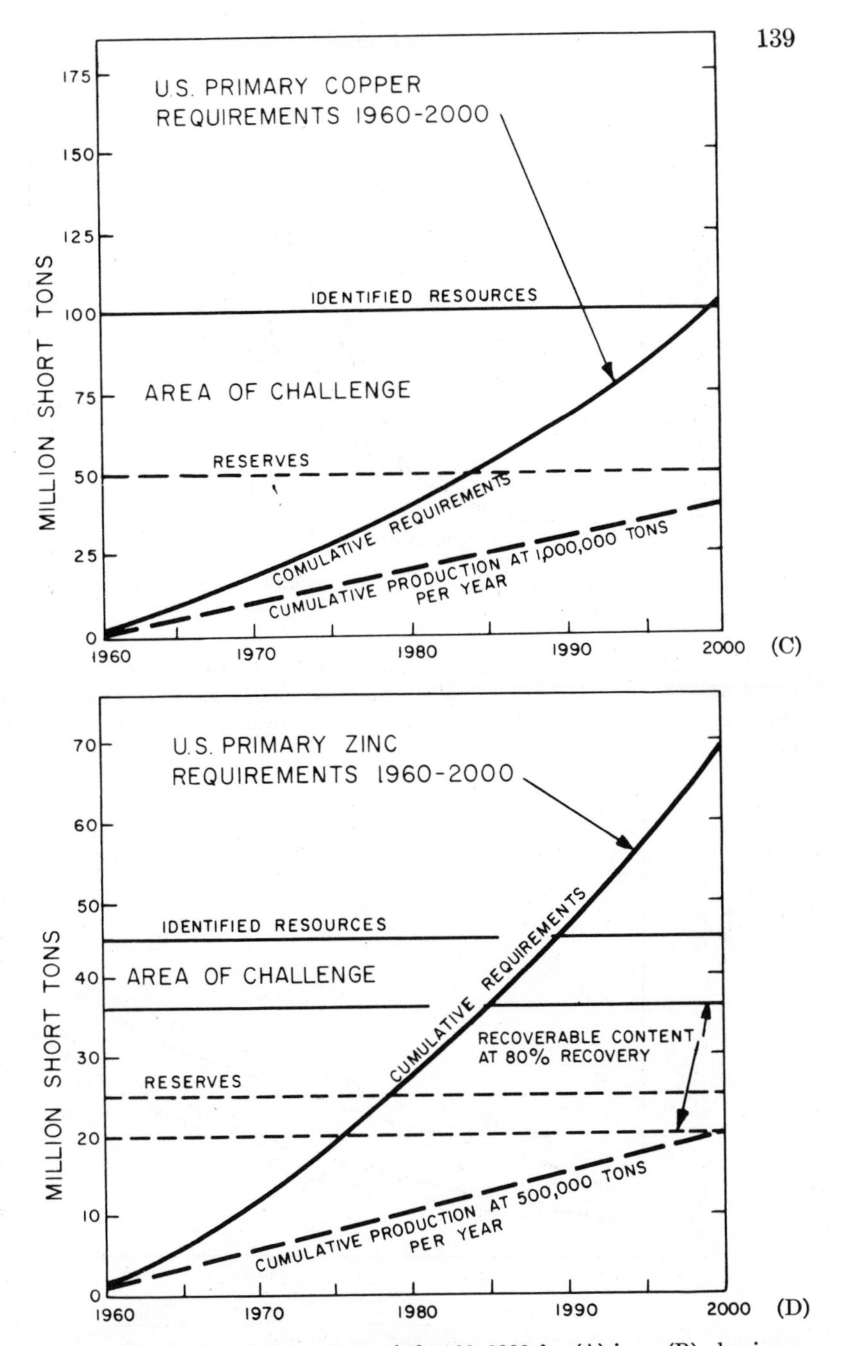

Figure 5.3. U.S. requirements for the period 1960–2000 for (A) iron, (B) aluminum, (C) copper, and (D) zinc. (Adapted by Engineering and Mining Journal[23] from "The Future Supply of Major Metals" by *W.* Netschert and H. Landsberg, published by Resources for the Future Inc., Washington, D.C., 1961).

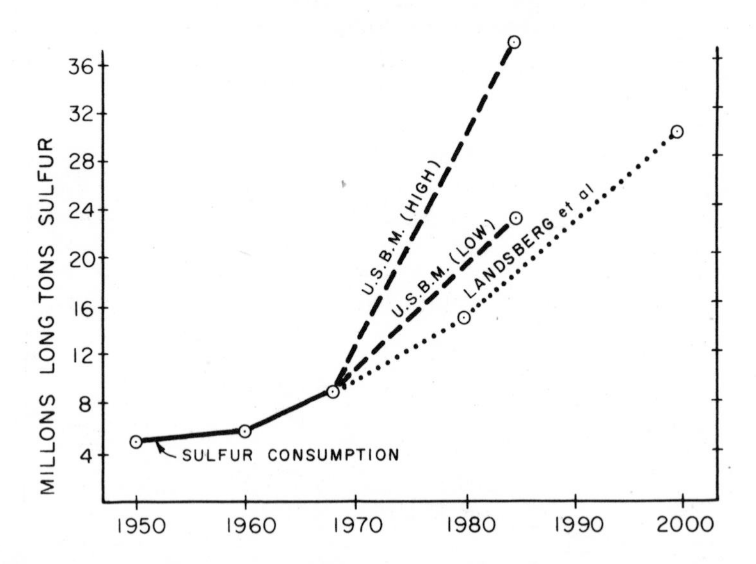

Figure 5.4. Current and future demand for sulfur. (After Proctor and Beveridge, *Population, Energy, Selected Mineral Raw Materials, and Personnel Demand, 2000 A.D.*, preprint no. 71-H-107, AIME, Centennial Annual Meeting, New York, N.Y., March 1971).

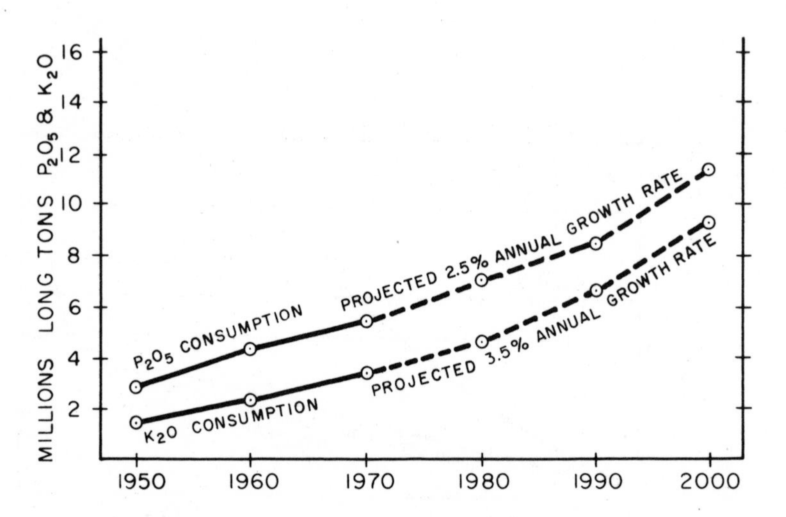

Figure 5.5. Current and future demand for phosphate and potash. (After Proctor and Beveridge, *Population, Energy, Selected Mineral Raw Materials, and Personnel Demand, 2000 A.D.*, preprint no. 71-H-107, AIME Centennial Annual Meeting, New York, N.Y., March, 1971).

and the burning of coal for power generation, would necessitate these industries to remove or to recover the valuable sulfur which will then be available to satisfy our future demands.

As far as the fertilizer minerals such as potash and phosphates are concerned, it is quite clear that the burgeoning world population is certain to tax the available supply of these essential commodities. Since unused lands with less fertility will be increasingly used in the future for our food needs, the projected annual growth rates in the United States of 2.5 percent for phosphates and 2.5 percent for potash appear to be rather on the low side. It is estimated that by the year 2000, more than 12 million long tons* of P_2O_5 content and 10 million long tons of K_2O content will be required for the U.S alone. In considering the world's needs, the current demand of about 60 million long tons of fertilizer is estimated to grow up to 110 million long tons by 1980, 220 million tons by 1990, and about 400 million tons by the year 2000.

Industrial Minerals

Industrial minerals are important resources for our growing industrialization and urbanization. These include building stone, crushed rock, sand and gravel, clays, gypsum, asbestos, limestone, dolomite, and the like. These minerals are used in large tonnage quantities and, after fuels, are the most economically valuable commodity. It should be noted that these resources have little value prior to their mining. However, after their removal and processing to useful materials they command a considerable increase in value. Thus, limestone and shale, the primary constituents of cement, may be worth only $1 per ton in the ground, but after their conversion to a useful product like cement, the price increases to about $20 per ton.

Energy Resources

In the preceding discussion we have seen that the availability of mineral resources in the future will depend to a large degree on the availability of the energy needed for their extraction from lower and lower grade deposits. Therefore, it is important that we examine the energy resources of the world to determine their adequacy. The prime energy sources for the present are fossil fuels (such as coal, oil, and natural gas) and water power (hydroelectric) which is comparatively a smaller source. Since fossil fuels are nonrenewable, their gradual depletion has caused considerable concern to the industry and government agencies. It is possible that additional

* One long ton = 2240 lbs. = 1.12 short tons.

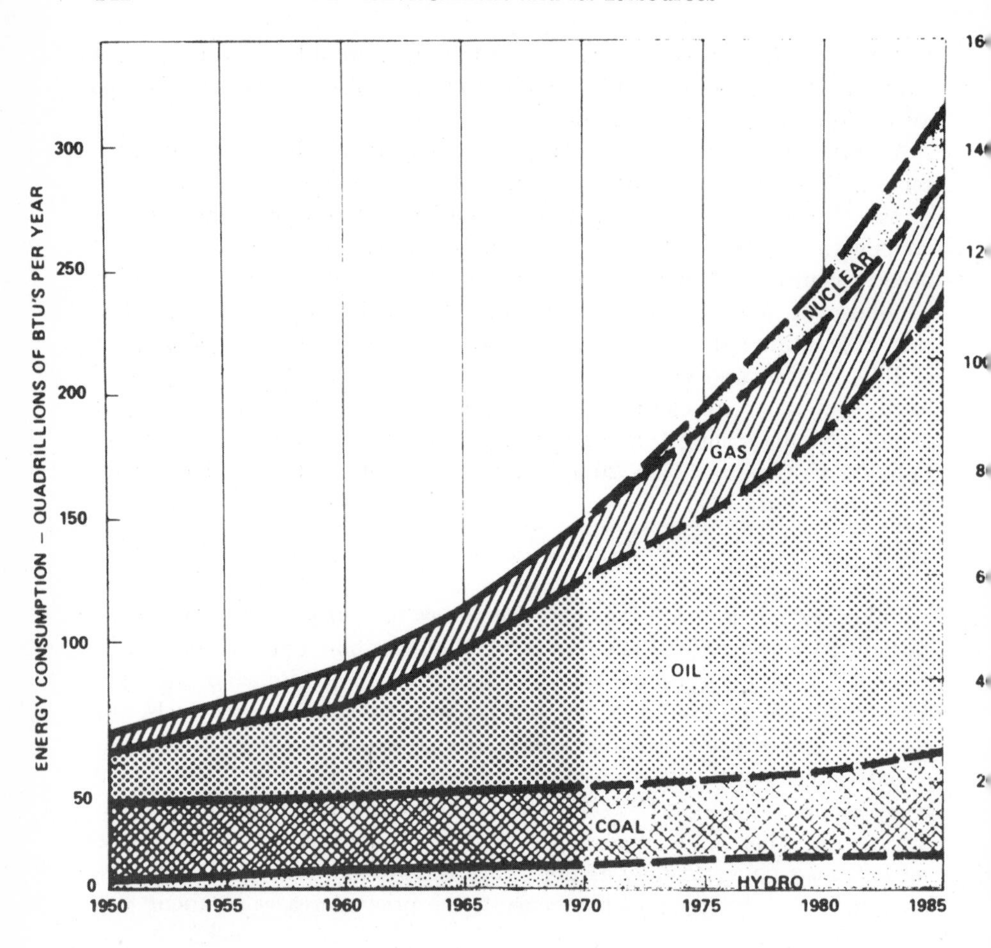

Figure 5.6. Free world energy consumption by fuel, 1950 to 1985. (After W. B. Davis, 1971, *World Fossil Fuel Economics*, "Centennial Volume", AIME, New York, N.Y.).

oil sources such as tar sands and oil shales will be exploited to obtain badly needed energy but again they will be limited in supply. However, it appears that nuclear energy, which today is in its infancy, is destined to be the major energy source of the future. It should be noted that fissioning of one pound of uranium produces heat energy equivalent to 5,900 barrels of crude oil. Besides nuclear energy, we have other sources such as solar and tidal energy, but these are not very likely to develop into major sources in the near future. Geothermal power may have a major impact in anomalously hot areas such as California, however, the impact will be of localized importance.

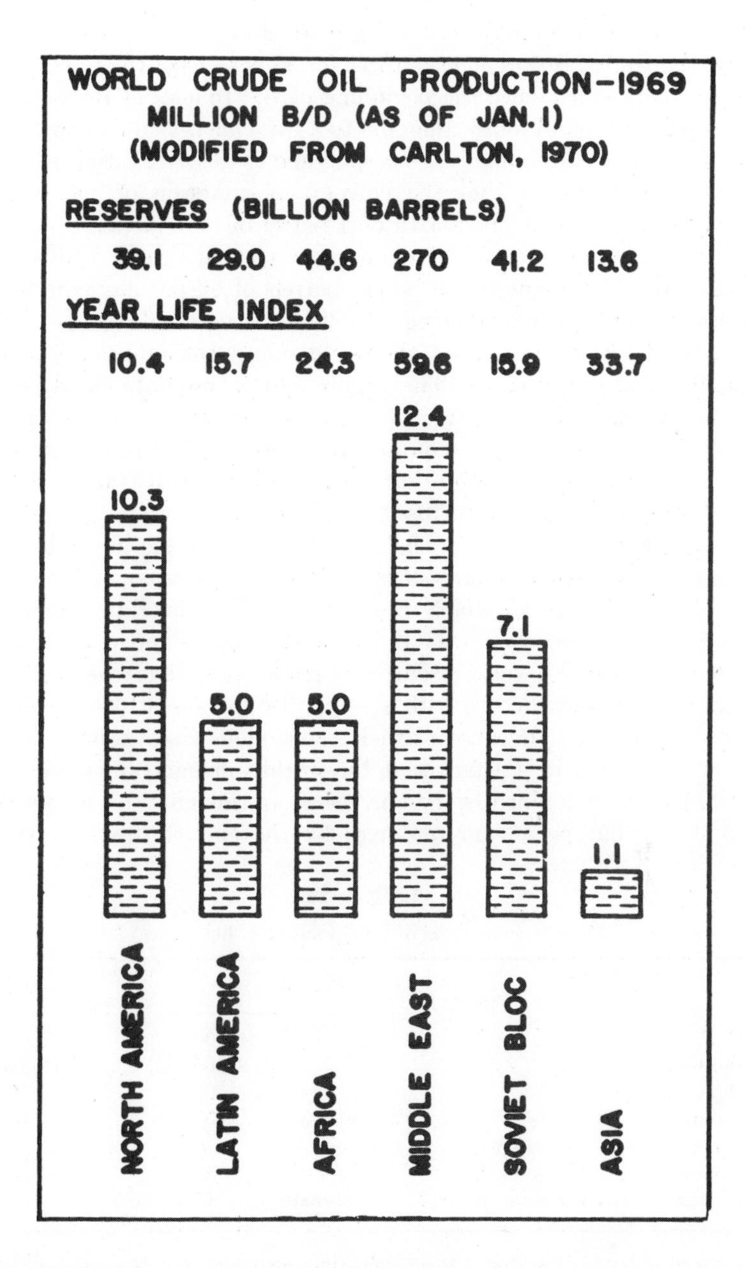

Figure 5.7. World crude oil production by producing areas in 1969 along with their known reserves and year life indices. (After Proctor and Beveridge, *Population, Energy, Selected Raw Materials, and Personnel Demand, 2000 A.D.*, preprint no. 71-H-107, AIME Centennial Annual Meeting, New York, N.Y., March, 1971).

The basic need for any nation's growth and progress is energy. Currently, the United States is responsible for 32 percent of the world's energy consumption, which since the beginning of the Industrial Revolution has risen consistently and today amounts to a total energy equivalent of about 9 billion tons of coal annually, corresponding to nearly 2.5 tons of coal per capita. If we consider that the current consumption of energy in the United States is about one-fourth of a barrel of oil equivalent per capita per day, then the total world requirements at this same United States standard would amount to 0.9 billion barrels of oil per day; the total requirements in the year 2000 being 1.7 billion barrels of oil per day.

In a recent study by Davis,[25] the world energy consumption is analyzed from 1950 and projected to 1985 (Figure 5.6). These data clearly indicate that oil and natural gas, have up to now, and are expected to supply two-thirds of the total energy requirements until 1985. From a long-range viewpoint, there is no doubt that nuclear energy will superseed all the other forms of energy, since fossil fuels are nonrenewable resources and have a projected "year life index" which represents the life of known reserves at the current production rate. This is illustrated by Proctor and Beveridge[24] in Figure 5.7 which shows the world crude oil production from various areas in 1969 along with their known reserves and year life indices. It is obvious that the world resources of crude oil are truly limited.

Skinner[26] has provided us with a comparison of fossil fuels resources as shown in Table 5.4. Because of their tremendous size we may feel complacent. However, it is a fact that the world's demand for crude oil and natural gas have doubled every ten years. At this growth rate we should be depleting our petroleum resources by the end of this century. For-

TABLE 5.4.

Potential resources of fossil fuels and their energy*

		Barrels $\times 10^9$	Btu $\times 10^{15}$
Coal		34,500	197,000
Oil and Gas		2,500	14,250
Oil Shale		1,000	5,700
Tar Sands	at least	600	3,400
1966 World Energy Consumption	equiv. to	30	170
1970 World Energy Consumption	equiv. to	36	198
1976 World Energy Consumption	(estimated)	48	270

* 1 barrel of crude oil is taken to be approximately equal to 0.22 tons of coal and to generate 5.7×10^6 Btu of energy.

Modified after B. J. Skinner, EARTH RESOURCES, (*C*) 1969. Reprinted by permission of Prentice-Hall, Inc., Englewood, Cliffs, N.J.

tunately, it is anticipated that the petroleum industry will not only discover new oil fields but will also learn to extract unrecoverable oil that still exists in known reservoirs. The next 30 years should provide sufficient time for man to develop other fossil fuel energy sources such as oil shales, tar sands, and coal. Brown, et al[2] have estimated that the world's available resources of coal may amount to as much as 6000 billion tons. Assuming a recovery of 50 percent for easily minable deposits, we arrive at a figure of about 3000 billion tons which should, at the current rate of consumption, last us for about 330 years or at the anticipated accelerated rate of the year 2000, for at least 160 years.

In looking briefly at the nuclear energy picture we find that as the world reserves of fuels become depleted around the middle of the next century, nuclear energy will be increasingly used from a projected figure of 6 percent in the year 1980 to about 25 percent by the year 2000, about 50 percent by 2050 and about 75 percent by 2100. Such a trend is clearly illustrated in Figure 5.8 which shows the possible pattern in the consumption of different energy sources projected into the next two centuries. As can be seen, fossil fuel consumption will reach a peak around 1975 after which it will level off until about 2030 and then decrease until it is completely depleted around 2075. The consumption of coal, on the other hand, will continue to increase but its use will be limited to the production of liquid fuels and chemicals. By the middle of the next century, most of our energy needs will be supplied by nuclear energy.[27]

If this is true, then do we have sufficient supply of uranium and thorium to sustain our growing industrial world? Since these sources are also nonrenewable, it is evident that their reserves are finite. However, the total energy available to us from uranium and thorium is considerably greater than the energy contained in our fossil fuels. This assumption is based on the fact that uranium and thorium are found in common rocks of the earth's crust in low but significant quantities. A typical granitic rock contains about 4 ppm uranium and about 12 ppm of thorium. These contents may seem insignificant, yet one ton of granite containing such small amounts of these elements possesses energy equivalent to 50 tons of coal. If we assume that two-thirds of this energy will be used up in extracting the contained uranium and thorium in granite, we can still count on one ton of granite to contain energy equivalent to about 15 tons of coal. Thus, man can always depend on this huge source of nuclear energy even when all the higher grade sources are exhausted.

In regard to higher grade sources of uranium and thorium, the U.S. Bureau of Mines has reported up-to-date reserves for the non-communist world.[28] The reserves, as shown in Table 5.5, are based on the price of uranium at $8 per pound of U_3O_8. At this price only the richest grades of

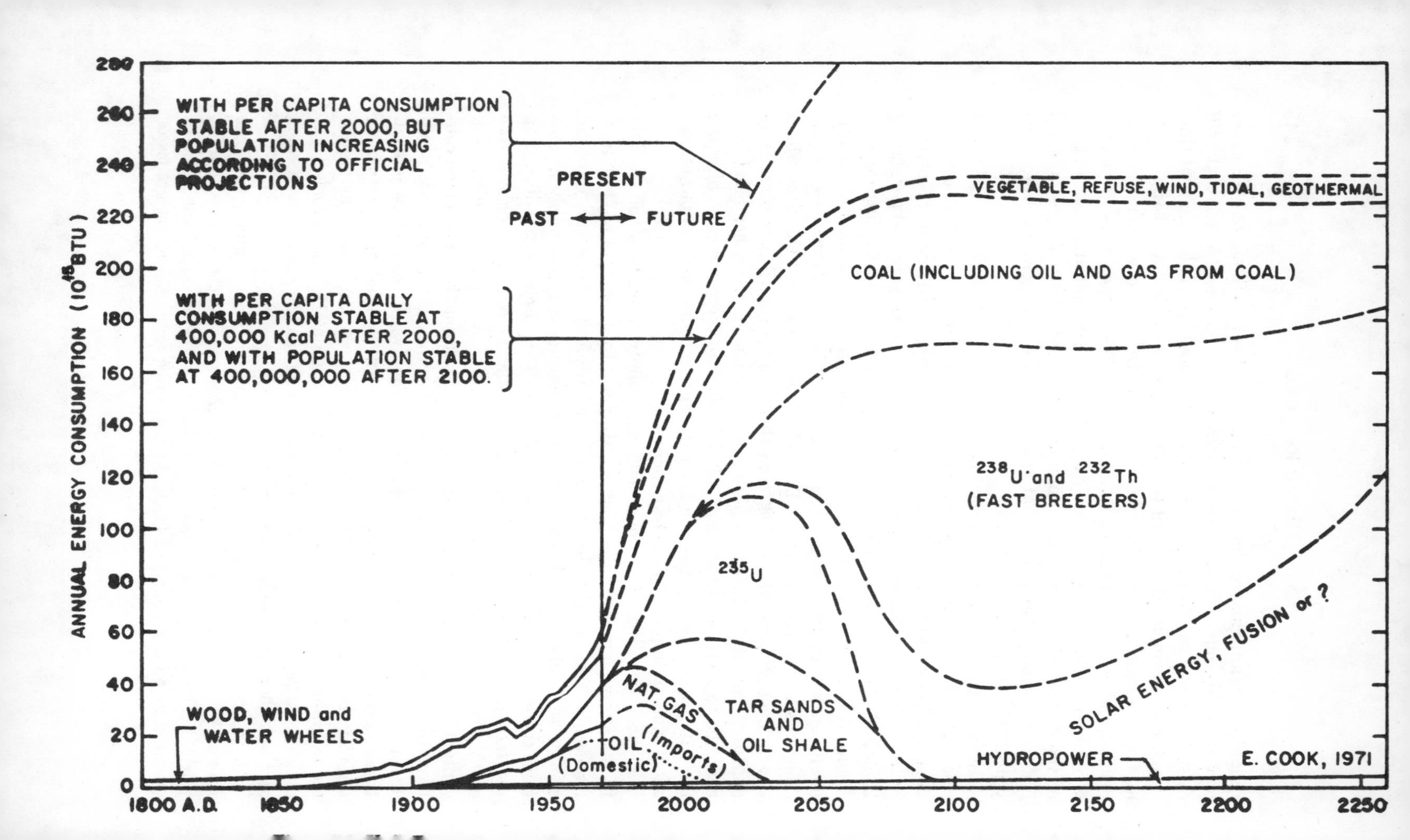
ANNUAL ENERGY CONSUMPTION (10^{18} BTU)
WITH PER CAPITA CONSUMPTION STABLE AFTER 2000, BUT POPULATION INCREASING ACCORDING TO OFFICIAL PROJECTIONS
PRESENT
PAST
FUTURE
WITH PER CAPITA DAILY CONSUMPTION STABLE AT 400,000 Kcal AFTER 2000, AND WITH POPULATION STABLE AT 400,000,000 AFTER 2100.
VEGETABLE, REFUSE, WIND, TIDAL, GEOTHERMAL
COAL (INCLUDING OIL AND GAS FROM COAL)
^{238}U and ^{232}Th (FAST BREEDERS)
^{235}U
SOLAR ENERGY, FUSION or ?
NAT. GAS
TAR SANDS AND OIL SHALE
OIL
(Imports)
(Domestic)
WOOD, WIND and WATER WHEELS
HYDROPOWER
E. COOK, 1971
1800 A.D.
1850
1900
1950
2000
2050
2100
2150
2200
2250
0
20
40
60
80
100
120
140
160
180
200
220
240
260
280

TABLE 5.5.

Estimated non-communist world reserves of uranium at price of $8 per pound*

(Thousand short tons of U_3O_8)

Country	1965	1970	
		Assured	Possible Additional
Argentina	5	9	21
Australia	15	11	3
Canada	210	200	290
Congo	6	6	NA
France	37	45	20
Niger	NA	12	13
Portugal	7	10	7
Spain	11	11	NA
Republic of S. Africa	140	205	15
United States	195	300	350
All others	16	17	24
Total	642	826	743

NA—Not available
* U.S. Bureau of mines, Minerals Yearbook 1965 and Minerals Facts and Problems, 1970.

ores (about 0.2% or higher U_3O_8 content) can be worked profitably. However, if the price of uranium increases as the result of improvements in the technology of breeder plants permitting the use of the abundant U^{238} isotope, it would be possible to work lower and lower grade uranium ores which fortunately are quite large. The United States Atomic Energy Commission realizing the relationship between the price of uranium and potential workable reserves, has estimated that at $10 per pound of U_3O_8, the U_3O_8 potential of the United States amounts to 650,000 tons. At $30 to $100 per pound U_3O_8, the reserves may be increased to 20 million tons; and at $100 to $500 per pound U_3O_8, the reserves would amount to about 1.7 billion tons. It would be interesting to note that 20 million tons of U_3O_8 at $30–$100 per pound U_3O_8, would generate energy equivalent to $250{,}000 \times 10^9$ barrels of crude oil, a quantity which is seven times larger than the total energy contained in the present fossil fuel supply of the world.

In regard to the cost of producing electricity by nuclear energy, it now appears that eventually it would be possible to do so for less than 10 mills (1 cent) per kilowatt-hour. A low figure of 4 mills was suggested at the International Conference on the Peaceful Use of Atomic Energy at Geneva in 1955. Presently, United States forecasts suggest a cost ranging from 4 mills to 6 mills per kilowatt-hour. In comparison, the cost of generating electricity using new coal-fired units in the United States in 1970, is about 6 to 7 mills per kilowatt-hour. Therefore, it is quite likely that in the not-too-distant future, nuclear electricity will compete with that obtained from coal-fired units.

In reference to nuclear power generation technology it should be noted that there are basically two types of reactors that have been considered for producing energy. The first one is the *fission* type using uranium as a fuel. This reactor derives its energy from the splitting of the uranium atom which in turn liberates neutrons that are capable of causing fission of other uranium nuclei. These can produce more neutrons which cause further fission and so on. The currently operating nuclear reactors are all of the fission type and from an environmental viewpoint are looked upon as dirty and waste-complicated. A newly modified version of the fission type is the *fast breeder* reactor. Such reactors have their own hazards but are relatively economical on fuel. These projected fast breeders use liquid metal, such as sodium, to transfer heat from the reactor to steam turbines. They are called breeders because they will make more fuel than they consume, thus relieving the present dependence on limited uranium reserves. Because of this inherent advantage breeder reactors are the preferred generators and in recent years have received much favorable attention and government research funds for their commercial development.

The second reactor is of the *fusion* type which derives its energy through a thermonuclear reaction in which nuclei of light atoms join to form nuclei of heavier atoms, such as the combination of deuterium atoms to form helium atoms. The enormous amount of energy produced in the stars, including the sun, has been attributed to nuclear fusion. One of the major advantages of this type of reactors is that it is a clean one resulting in a minimum deterioration of the environment. Furthermore, if energy could be obtained by the fusion of hydrogen nuclei, then the water of the ocean would constitute a tremendous reservoir of energy. For these reasons, it is obvious that fusion reactors are destined to be the leading energy producers of the future.

In summarizing, we can confidently say that man has sufficient potential sources of energy to last him for a very long time. It is also clear that our future energy problems will be related to technology rather than to lack of resources and the development of new energy sources will depend to a large degree on political, economic, and environmental factors.

5.4 ECONOMICS OF MINERAL RESOURCES

Significance of the Mineral Industry

Throughout history minerals have been important to humanity. Man's material progress is measured by his use of mineral resources. These pervade every sector of an industrialized society and every evidence indicates increased dependence on minerals in the future economy. In the last 30 years, the United States has consumed more minerals than the rest of the world since the beginning of civilization. In the next 30 years, the demand for minerals in the nation is expected to be three to five times the current level. This tremendous increased in demand will be primarily due to continuing population and economic growth as well as to technological progress and innovations in the extraction and utilization of minerals. There is no doubt that abundant and varied mineral supplies have enabled us to sustain a diversified industrialization and as a result, the United States enjoys the largest *gross national product* (GNP) and one of the highest standards of living in the world.

The gross national product, a popular measure of economic growth of a

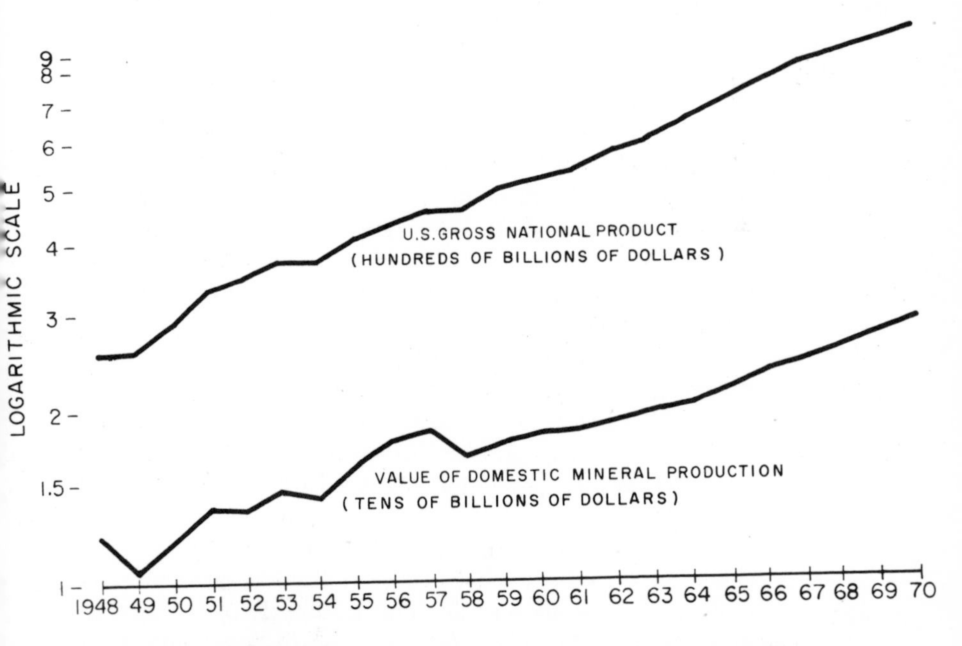

Figure 5.9. The U.S. gross national product and the value of domestic mineral production (Updated from "Mineral Science and Technology—Needs, Challenges and Opportunities" National Academy of Sciences, Washington, D.C. 1969).

nation, is based on population growth multiplied by buying power and represents the total national output of goods and services at existing prices. Since the buying power in an industrial economy is proportional to the availability and utilization of minerals, the gross national product is directly related to the growth of the nation's minerals industry as shown in Figure 5.9. It is interesting to note that both the GNP and the mineral industry are growing at a comparable rate of about 4 percent per annum. Recent projections of mineral requirements until the year 2000 by the United States Bureau of Mines[29] indicate that the growth of domestic demand for minerals during the next thirty years will continue to range from 3.5 to 5.0 percent per year. Thus, looking towards the future, mineral resources are destined to play an even more important part in man's progress than in the past.

The collateral aspects of minerals and the industries based on them are important elements in a nation's economy. This essential activity constitutes a source of income, employment, tax revenue, and cargo in national and international trade. Minerals also constitute a base for economic, political, and military power. Exploitation of mineral resources provides the basic income on which many other activities depend. However, this fact has been overlooked in many instances and it is only when a prosperous mining community turns into a ghost town on termination of mining operations that the significant impact of this activity on the local economy is realized. Such cognizance is also illustrated when the miners go on strike. Rapid development of Venezuela and of the smaller countries adjoining the Persian Gulf have been profoundly affected by income derived from the exploitation of their oil resources. On the other hand, the recent decline of national income from tin and silver in Bolivia has led to its economic difficulty and political instability. However, it is entirely possible that the improvement in the Bolivian situation may be brought about by the development of other mineral resources.

In considering the economics of mineral resources it is not necessary that the industrial growth and economy of a country be based solely on the existence of the mineral resources in that country. Some nations such as Japan, Great Britain, Germany and others do not possess extensive mineral resources. However, they obtain the needed resources from other countries and their economic and industrial vigor functions as a catalyst to manufacture and distribute the products needed by people throughout the world. The increased mining activity in recent years in Australia and Canada as well as in some developing countries like India, is attributed to the increasing demand for raw materials by Japan (see Figure 5.10).[30]

In the following chapter, we will see that metals are quite durable and many of these can be reclaimed and reused many times. The maintenance

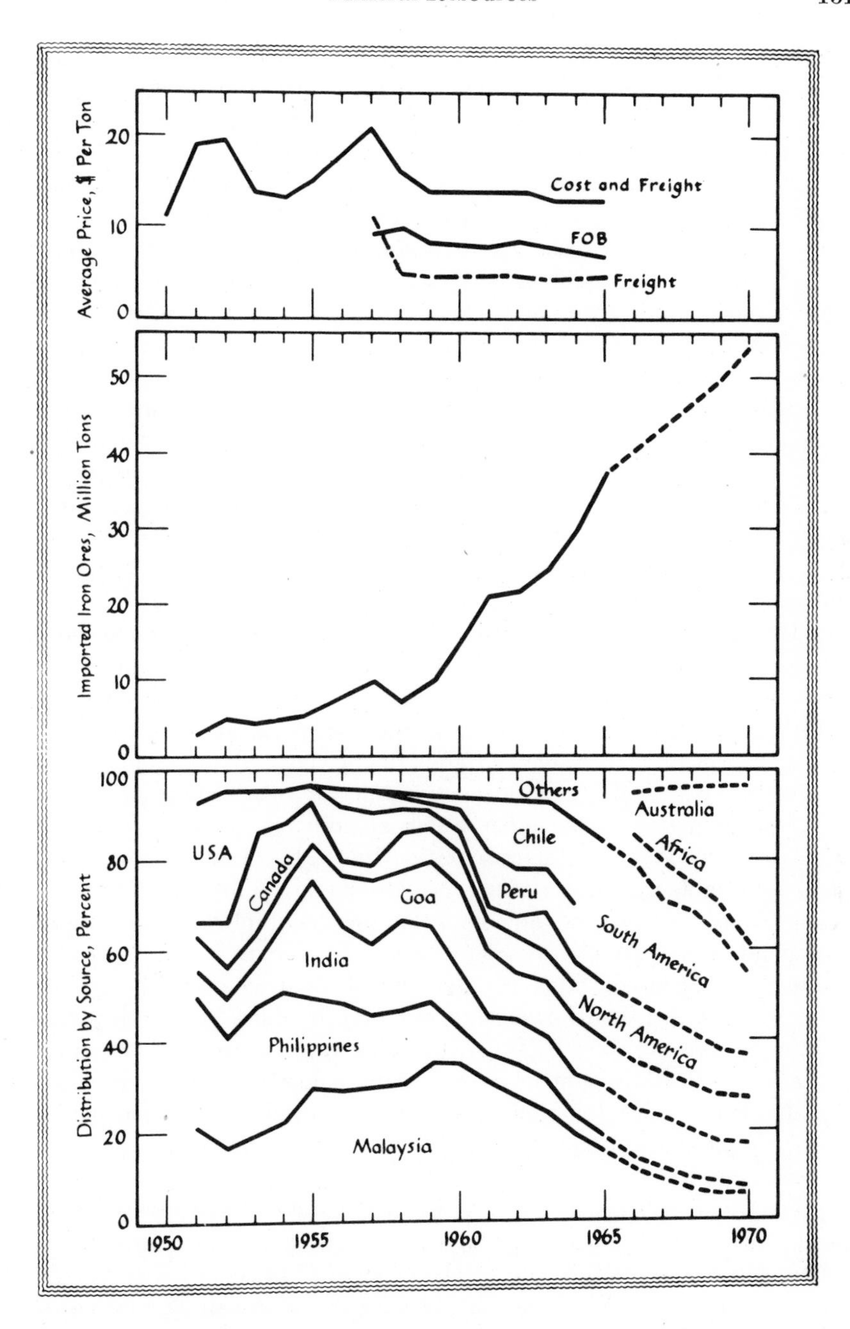

Figure 5.10. Trends of the amount, price and percentage by source of iron ore available to Japan. (After Tanabe, Takahaski, and Iwasaki, 1967[30]).

of such *minerals-in-use* depends on the period of useful service and their reclamation for future service. Some uses allow virtually complete recovery of minerals while others consume the commodities beyond recovery. Thus, a metal like gold can be made to serve indefinitely while mineral fuels after their initial combustion cannot be reused as a source of energy. Likewise, lead and titanium used as paint pigment and fertilizers applied to soil are not available for reuse. However, in such cases, the period of useful service is to be noted. The mineral pigments or construction materials such as cement and stone, though not reusable, remain in service for a long time while the fertilizer mineral is available to the soil only for a short time. This is also the case when steel is used for constructing a building which will be in use for fifty to a hundred years or it is used in an automobile and has an average cycle of use of only twelve years.

With the rise of our industrialized economy and with the importance of minerals in military affairs, assurance of access to mineral supplies has become more urgent. Today, more than ever before, great importance is attached to having adequate and dependable mineral supplies in industrial planning. From the consumer's point of view, national self-sufficiency in all minerals would be an ideal situation. Such happy circumstances, however, do not exist in any nation and the hazards of dependence on foreign sources for minerals in short supply may be reduced by enlarging and stabilizing international trade, tariff or quota systems, and stockpiling.

The status of mineral property rights, in the economy of a country is a controversial issue. Many nations still practice the regalian theory of subsurface property rights, under which all mineral resources belong to the crown or the state. In this case, either the state operates the mines or sets up a suitable system to induce private parties to operate the mines. In other countries, the governments retain ownership of subsurface rights but depend on private operations to work the mines under concession or lease. In tne United States, certain mineral resources on public domain are not subject to private ownership but may be mined privately under the Mineral Leasing Act. However, the idea that has proved extremely stimulating to the development of the mining industry is the one in which the individual retains the rights to subsurface property. Under such a system, the economy has been rapidly enriched because it has provided maximum incentives to the operators. The only objectionable aspect of the private ownership is that it is considered antisocial and monopolistic, and in some cases, the operators have exploited the resource without regard to ultimate extraction and at the expense of degradation of the environment. However, it is believed that alleged abuses can be reduced by effective regulations without impairing private incentive.

Distinctive Features of Mineral Economics

Some of the important features peculiar to mineral resources and thus to their economics are:

(1) *Localized occurrence* which means that the mineral resources of a nation are fixed in a certain locality and they must be worked where they occur. These mineral treasures constitute a temporary asset to the countries possessing them and give them great commercial advantage over less fortunate trade rivals.
(2) However, due to *exhaustibility* of mineral deposits, this prosperity is short-lived since each individual deposit has its limits and if worked long enough, sooner or later it is depleted. Because of the limited assets, all the capital invested in a mine should be paid back with a profit before the deposit is exhausted. In mineral economics this is accomplished by amortization, namely, the liquidation of a debt by means of a sinking fund.
(3) *Increase of costs with depth* is a common feature of all mining as well as oil and gas production operations.
(4) *Discovery hazards* are always associated with the discovery of hidden mineral deposits. This activity requires a great deal of mining skill and even luck. A new "strike" (discovery) may bring rewards to the finder but may put submarginal producers out of business.
(5) *Expendable minerals* are those commodities that can be used only once since they are either destroyed or converted into an unrecoverable form. Such minerals include coal, petroleum and mineral pigments.
(6) *Non-expendable minerals* on the other hand, include metals such as gold, copper, and lead which can be reused many times. Articles made of such metals are recycled as scrap and it is obvious that recycling of scrap and reuse of metals will become a very important activity in the pollution conscious world of the future.

Besides these peculiar features of mineral economics there are some economic factors that are common to mineral as well as manufacturing industries. These factors influence marketing and production costs and include items such as demand, supply, cartels, substitutes, market speculation, and production costs. The *demand* for minerals and metals is directly related to the activities of the manufacturing industries. A reserve stockpile is always provided for the smooth operation of plants. Such stockpiles are an asset in time of increasing business volume but are a liability in times of depression. *Supply* of metals controls the price of the commodity. This is expecially true if the oversupply of metal results from its production as a by-product. Scrap, too, is an important source of metal supply and as pointed out earlier, this source is destined to play a very important role in the future.

Because of the uncertainties of the metal market and its price structure, metal producers and marketers resort to *cartels* which are written agreements to regulate production or sale so as to maintain a certain price. Such agreements are usually workable only over a short time and that, too, not very effectively. However in case of aluminum and tin, such cartels have proved extremely beneficial in controlling their prices. A cartel can also be in the form of an agreement between governments.

Satisfactory *substitutes* are readily available for almost all raw materials, if the price is not a prime consideration. The search for a satisfactory substitute is justified whenever the price of a commodity is controlled at a high level or there appears to be a danger of a scarcity developing. *Market speculation* affects the price of minerals and metals just as it does other commodities.

Finally, *production costs* in the mining business are figured on the same basis as for any other industrial activity. Such costs depend largely on capital costs, maintenance costs, operating costs, and amortization.

Importance of Minerals to the Economy of a Nation

We have already seen that mineral resources are essential for an industrialized civilization and that they contribute significantly to the wealth and power of a nation. Since mineral resources constitute not only wealth but the basic ingredients of modern civilization, they profoundly influence the economics and politics of a nation. In the past, nations have for centuries fought wars to gain possession of important minerals that they lacked in their own territories and have likewise fought wars to protect the trade routes through which their badly needed minerals were supplied. The essential minerals in which a nation is deficient are commonly referred to as the *war minerals* which are further divided into *strategic, critical* and *essential.*

Strategic minerals are those commodities that are needed by the industry but are not produced at home in sufficient quantities to meet normal industrial demands. In normal times, these minerals are supplied by foreign sources and the list differs for different countries and changes from year to year for individual nations.

Critical minerals, on the other hand, are those necessary for the industry and which are normally produced in too small a quantity to satisfy the domestic demand; but known underdeveloped supplies are available and are sufficient to satisfy the nation's need during an emergency. In this case there is no exportable surplus.

The *essential* minerals are those most vital to industry and which are produced in sufficient quantity to yield an exportable surplus. Such minerals are of sufficient military interest to justify attention, and official regulation is necessary. For a detailed coverage of the three types of war minerals, readers are referred to specific references[31,32].

In order to provide adequate supplies of needed raw materials for the production of military weapons and items necessary for civilian defense, the government resorts to *stockpiling*. This is a program of accumulating government-owned supplies or inventories of basic raw materials. These materials represent commodities whose domestic production is not sufficient to supply needed requirements for defense purposes. The government therefore, obtains the required resources from friendly foreign countries and holds them until they are needed. Stockpiles are more or less a form of insurance designed to allow the economy of a nation to function in the event the foreign sources of supply are cut-off. Because of the great significance of stockpiles in national defense as well as economy, their accumulation, maintenance, and release must be given careful consideration.

Minerals in National and International Affairs

The exploitation of mineral deposits engenders many unique problems in national as well as international affairs because of their evanescent nature, restricted occurrence, and indispensability to industry throughout the world. Mineral availability has been instrumental in determining the course of history in the past and is destined to be of increasing importance in the future. The importance of mineral resources to the United States' welfare was firmly established during both World Wars. Past experience has also shown that in times of peace as well as in times of war certain mineral resources are indispensable to an industrialized nation.

If we examine the principles related to the national and international control of minerals, we find that the following, elemental considerations are basic to the formulation of policies affecting mineral resources:

(1) No industrialized nation has ever been self-sufficient in all the mineral resources. Therefore, international movement of minerals is necessary.
(2) International movement of specific minerals cannot be stopped by enactment. Of course, it may be aided or hindered locally and temporarily by tariffs, and embargoes. However, measures to

permanently close the main international channels created by nature are doomed to failure.

(3) In order to provide efficient transportation, minerals should be concentrated, refined, and fabricated at or near the source of supply whenever favorable conditions prevail.

(4) Continuous exploration and development of mineral resources are essential in order to maintain adequate supplies of minerals. Restrictions which interfere with the necessary searching efforts are undesirable and should be avoided, except under conditions of national crisis. This implies an open-door policy, national and international, under which there are no restrictions of the issuing of mining licenses and concessions to foreigners. The right of nations to control their own mineral resources in times of war, is of course, unquestioned.

(5) Developing nations possessing important mineral supplies needed by the world should be aided to develop their resources through international efforts or through other bilateral agreements between governments, so that there will be equal opportunity for all nations.

As mentioned earlier, the irreplaceable nature of mineral deposits and their unpredictable occurrences are the two important features of mineral geopolitics. These factors play an important role in the internal and external affairs of nations. In international mining activity, the operating profit-investment ratio is highly important and nations seeking foreign investment in the development of their resources must create a favorable investment climate that adequately provides for depletion and high rate of return on the monies invested. On the other hand, a nation's security in mineral raw materials is adversely affected by depletion. England and Western Europe have already experienced the consequence of mineral exhaustion, especially in nonferrous metals. The United States, too, is feeling the pinch of depletion by shifting from virtually self-sufficiency in fluorspar, lead, zinc, iron ore, and petroleum to substantial dependence on foreign sources. Russia, on the other hand, contains a vast virgin mineral wealth which is yet to be developed and in this case a depletion factor is no deterrent to industrial expansion.

From the international viewpoint, the unequal pattern nature has followed in depositing minerals is primarily responsible for determining the centers of industrial and political power in the modern world (see Figures 5.11 and 5.12).

The unequal distribution of mineral wealth is also responsible for large international trade in mineral commodities. Many times, the "have not"

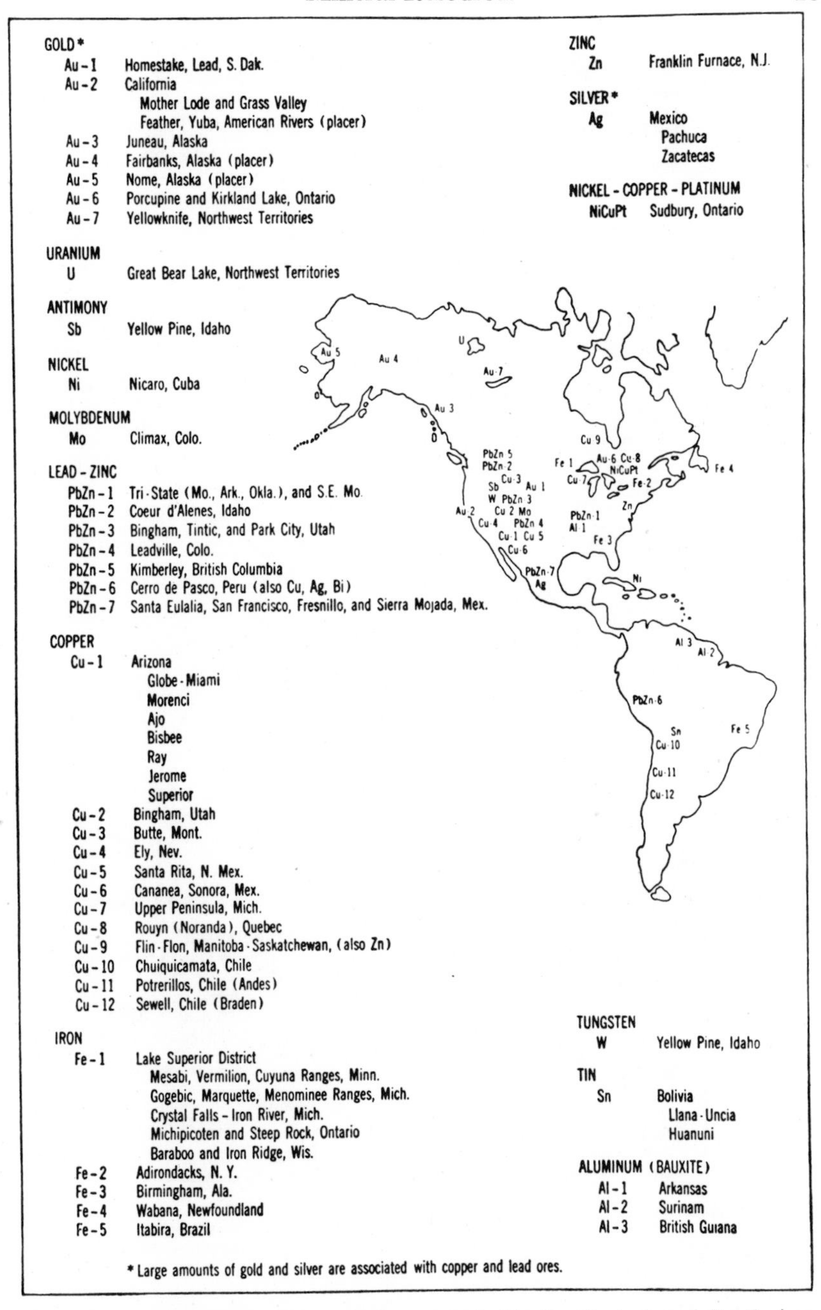

Figure 5.11. Important metal-producing areas in North, Central, and South America. (After J. Newton, "Introduction to Metallurgy," © 1948. *Printed by permission of John Wiley and Sons, Inc., New York, N.Y.*)

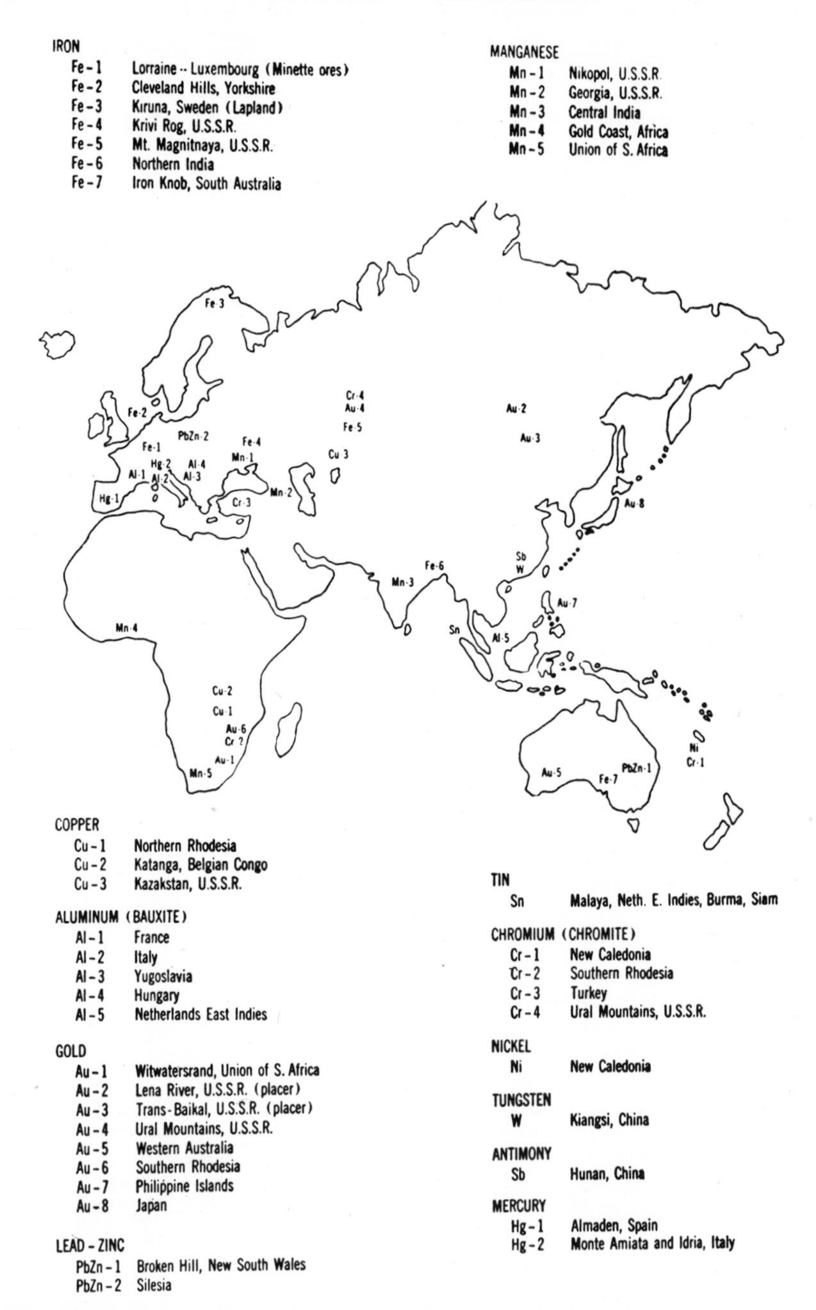

Figure 5.12. Important metal-producing areas in Europe, Asia, Africa, and Australia, (After J. Newton, "Introduction to Metallurgy" © 1948. *Printed by permission of John Wiley and Sons, Inc., New York, N.Y.*).

nations in order to improve their mineral security have resorted to economic autarchy, conquest, and ultimately to war. In some cases, nature has created concentrations of essential minerals in single deposits thus favoring the establishment of a monopoly.

During recent times, some of the developing nations have sought to profit from their country's mineral resources by nationalization, that is, restricting exploitation of raw materials to nationally controlled companies (private or state owned). However, such methods have proven shortsighted from most nations' own point of view. In some instances, the politicians have taken advantage of the situation to gain power.

Nationalization of the mining industry in an underdeveloped country usually overlooks the necessity for financing, developing, maintaining and operating the mines, mills, smelters and refineries. A strict nationalization policy applied over many years to most minerals and the preclusion of private development have actually increased the poverty in what is potentially a wealthy nation and have resulted in out-of-hand inflation. By all means, there are some positive and beneficial aspects of nationalization such as development of their own talents, self-sufficiency during times of international stress, and improved educational facilities. Moreover, it prevents the exploitation of the nation's valuable resources by unscrupulous foreign entrepreneurs. Nevertheless, it is a proven fact that extreme nationalism is not healthy since it prevents a nation from participating in and profiting from international trade. It is hoped that in the future both extremes will be avoided and that underdeveloped nations will seek a workable foreign capital-local participation system of benefit to all parties concerned.

Modern geopolitical philosophy emphasizes industrial power and acknowledges the importance of mineral resources as a means of achieving such power. Leith[33] was the first American to recognize the vitality of mineral resources in political and military strategy. He stressed the international problems arising from the unequal distribution of mineral wealth, the need for orderly and equitable apportionment of available peacetime supplies among the nations of the world, and the relationship of minerals to world peace and world power. Pehrson[34] analyzed power politics and mineral resources on a hemisphere basis.

Our society is heavily dependent on the internal combustion engine. Thus, assurance of adequate petroleum supplies is a necessity, but possession of oil resources by itself does not guarantee industrial development. A classical example is that of the Middle East which in spite of the vast petroleum riches has not become an industrial power probably due to its lack of coal, iron, and other types of mineral wealth.

Man's material progress over the ages has been largely a result of technological change and further advancement is inevitable. The commercial development of nuclear energy is certain to affect national and international affairs profoundly but its alteration of the geopolitical importance of minerals will only be moderate. Since fissionable materials are themselves products of the mineral kingdom, man's basic dependence on minerals will not change. However, it will lessen the importance of fossil fuels as sources of energy and may provide an economical source of energy for industrial progress in developing countries. Atomic energy cannot displace coal as a reducing agent in the production of steel, but the required quantity of coal as a metallurgical reagent might conceivably in the future be transported to the iron-rich, coal-poor countries for the economic production of steel.

A National Mineral Policy

An adequate, dependable, and continuing supply of mineral resources is indispensable to a nation and its industries in meeting the needs of an expanding population, a rising standard of living, and the national security. It is the responsibility of a government to look ahead, anticipate future needs, and plan its activities to allow for the maximum contingency in mineral requirements of the nation as a whole. The government should also help further long-range research where the hope of immediate commercial application is not attractive to industry and to assist industry to solve short-range problems. Pursuit of these responsibilities requires continuing review of the national position regarding every mineral of significance and dissemination of comprehensive statistical data to other government agencies and to the public.

The above mentioned efforts of the government should constitute part of an overall *national mineral policy*. It is understandable that every country must have a policy which is consistent with the needs of their people and their available mineral resources. According to Park[35] an enlightened national mineral policy would be based on the following principles:

(1) It should consider the welfare of the nation as a whole;
(2) It should consider the special needs of the mineral industries;
(3) It should coordinate federal and state laws pertaining to minerals; and
(4) It should encourage international trade in minerals.

In order to provide the best climate for optimum utilization of mineral resources, industry and government should cooperate at all levels of effort to achieve the following objectives:

(1) Wise production and utilization of mineral resources;

(2) Discovery and development of new sources of minerals;
(3) Maintenance of adequate mineral reserves and stocks;
(4) Effective exploitation of domestic mineral resources consonant with fully foreseen requirements; and
(5) Assurance of access to foreign mineral markets to supplement domestic production as needed.

The above objectives can be attained by a realistic program based on the following guidelines:

(1) Appraisal of a country's mineral position;
(2) Development of submarginal resources;
(3) Possible new or wider uses for abundant resources;
(4) Development of substitutes; and
(5) Application of conservation.

Since a wise and efficient use of mineral resources depends to a large degree on advances in technology, the government has a unique role in coordinating and integrating research in the minerals industry. Such efforts should range from basic or pure research through applied technology and economic evaluation. These studies should distinguish between technical feasibility without regard to economic feasibility (research) and commercial applicability (engineering evaluation). Such studies should be followed by mineral-industry studies consisting of pilot plant, demonstration plant, and design of practical installation, along with market studies and economic evaluation. To be most useful and in the utmost public interest, such government research should not be competitive with other efforts and should concern itself with national needs and goals.

Wise Use of Mineral Resources

Some conflict of economic interest is always involved in the exploitation of mineral resources. Satisfying the present needs of minerals may deprive those with more dire need in the future. The excessive utilization of precious mineral resources in times of war may reduce critical civilian supplies needed in times of peace. Moreover, mining may degrade land surfaces that are suitable for agriculture, ranching, or urban development and water necessary for mining and extraction may be diverted from more useful purposes only to be returned contaminated for additional uses. Wasteful exploitation of mineral resources may occur when resources are large compared with current demands and competition among producers may force them to adopt wasteful low-cost methods.

Catering to immediate needs for metals and minerals does not necessarily lead to anticonservation practices. In many cases, the primary shoestring operation has yielded sufficient profits to finance large-scale operations for the economic exploitation of lower grade ores left behind. Furthermore, the experience and knowledge gained in the first operations have been indispensable to the successful working of the latter. Moreover, there is no guarantee that the future needs of a given commodity will be greater than present ones. For these reasons, it is essential that in planning for the availability of minerals in the future due emphasis be placed on the possibility of shift in demand due to more efficient utilization and substitution as well as on the advancements in minerals' extraction technology.

Opposing interests are exhibited by different producers exploiting a particular mineral. The producer with large reserves will act in a more statesman-like and conservative manner than a mine operator with limited reserves who is bent upon making a quick profit without regard to the welfare of the industry and environment. In many cases, the owners or stockholders are only interested in obtaining immediate high-income rather than a larger total income over a deferred period. Developing countries also experience the same dilemma since in the early stages of industrialization they would be interested in quick exploitation of mineral resources to obtain essential funds for various priority projects and programs. However, after industrial advancement, they are more conservative towards further exploitation of their dwindling natural resources.

One of the traditional areas of conflict involves the regulation concerning the private use of natural resources as opposed to their public utilization and management. Thus, the initiation of mining activities in and around predominantly agrarian communities results in many conflicts of interest. Strip mining and petroleum production activity require substantial withdrawals of agricultural or forest land. Surface use is also adversely affected by smelter smoke, tailings, slag piles, and the like. A more serious problem concerns the use of water by mineral processors as a source of power, in milling, and as a medium for discarding tailings. Especially, in many water-short areas the same water is needed for farming, municipal and recreational uses; and most of these users do not wish to use polluted water.

In recent years, multiple use of natural resources has received increasing attention both by government agencies and concerned citizens. It is now generally accepted that multiple use can be achieved without conflict of interest between different parties wanting to use the same natural resource. Wherever necessary, zoning laws and regulations prohibit or limit mining activity as is frequently the case in urban areas. Particularly in public domain, the problem of land management becomes quite complex since all conceivable uses are given serious consideration. However, judicious laws such as the Homestead Act, Mineral Leasing Act, and other mining laws make it possible to exploit our mineral resources and to obtain the needed metal, minerals, and fuels for our industrialized society.

Marketing of Metals and Minerals

The marketing of metals, minerals, petroleum and natural gas is a fascinating trade, international in character, sensitive to slight changes in economic conditions, and complex in its organization. Each individual commodity has its own marketing peculiarities and problems, the differences being a matter of tradition more than of necessity. However, in most cases, the differences are largely due to the very nature of the problems of producing or consuming the commodity.

Ores and concentrates are marketable products. In some cases they are treated near the mine with the milling and smelting done by the parent company. In other cases, the ore or concentrate is shipped a considerable distance and it may pass through several hands before finally reaching the consumer who extracts the metal. In some custom smelters, the ore is treated for a certain fee or *smelter charge* and then returned to the original owner of the ore. In most cases, however, the ore or concentrate is purchased outright and the material becomes the property of the smelter.

As we shall see in the section covering metallurgical extraction (5.8), in the case of several non-ferrous metals, as also with the precious metals, it is beneficial to *concentrate* the ore into a smaller bulk of such composition as gives maximum monetary yield when the buying schedule, freight, tailing loss, and milling cost are taken into account collectively. When the markets are sluggish or unsettled, both the *smelters* and *custom mills* may refuse to purchase, but they will treat the product at an agreed per-ton charge, either on a consignment basis to sell whenever possible, or, will agree to return the concentrate to the miner, who must then find his own market. In these cases, the *custom milling charges* are based on the kind of ore, the size of shipment, the extent of separation required and the degree of competition.

On the other hand, smelters have a more difficult problem in the purchase of base-metal concentrates or ores and the schedules of charges and the methods of payment are very complicated. Usually, the smelter buys the material on the basis of the assays, paying for the contained values at prices existing in principal metal-market centers (New York, London, etc.), either at date of purchase or at some agreed date of sale, less a charge covering the cost of treatment and profit thereon. The treatment charges include the cost of transportation, sampling, smelting and refining, selling, and a carrying charge of metal from time of purchase to the time of disposal.

TABLE 5.6

A typical smelter schedule (courtesy Joy Manufacturing Company, Denver Equipment Division)

Schedule	Copper smelter	Lead smelter	Zinc smelter	Crude lead-zinc milling ore
Treatment charge	$12.50 per dry ton	$12.50 per dry ton on lead content 25% wet assay plus 10¢ per ton for each unit under 25%.	$51.50 per dry ton based on 12¢ market price. Add $1.25 per net ton for each 1¢ or fraction above 12¢.	$4.00 + per dry ton.
Payments				
Gold	If .03 Troy oz. per ton or over to 5 oz. pay $31.8183 per oz.; $32.3183 per oz. from 5 oz. to 10 oz. and $32.3183 for 10 oz. or more.	If .02 Troy oz. per ton or over to 5 oz. pay $31.8183 per oz.; $32.3183 per oz. over 5 oz. to 10 oz. and $32.8183 over 10 oz. or more.	If .03 Troy oz. per ton or over pay for 70% at $34.9125 per oz.	If .02 Troy oz. per ton or over pay for 65% at $34.9125 per oz.
Silver	If 1 Troy oz. or more pay 95% of market price less 1¢ per oz.	If 1 Troy oz. or more pay 95% of market price less 1¢ per oz.	If 1 oz. or over pay for 70% at market price.	If 1 Troy oz. or more pay for 70% at market price.
Copper	Deduct 20 lbs. copper per ton and pay for 100% at market price less deduction of 2.75 cents per lb.	Deduct 20 lbs. copper per ton and pay for 95% at market price less deduction of 10¢ per lb.	If 1% or more copper pay for 65% at market price less deduction of 8¢ per lb.	If 2% or over pay for 70% of sulfide lead and copper combined at market price for lead less 3.27¢ per pound.
Lead	Pay for 50% at market price less 3.5 cents per lb.	Deduct 30 lbs. lead per ton from wet assay and pay for 90% of balance at market price less 2.77¢ per lb.	Deduct 30 lbs. lead per ton from wet assay and pay for 65% at market price less 2¢ per lb.	

Zinc			If 40% or over pay for 85% at market price less .583¢ per lb. If under 40% deduct 80 lbs. zinc per ton and pay as above.	If 2% or over pay for 100% sulfide zinc at 70¢ per unit on 12¢ market price. Add or deduct 6¢ per unit for each 1¢ change in market price.
Cadmium			Pay for 30% at 75% of price.	
Lime		Pay 5¢ per unit if 5% or more.		
Deductions				
Moisture		5¢ per unit over 10%.	1% minimum deduction from total weight.	
Zinc	30¢ per unit over 6%.	30¢ per unit over 8%.		
Iron			50¢ per unit over 6% Fe + Mn combined.	
Lime & Magnesia			$1.00 per unit over 1% combined.	
Sulphur		25¢ per unit over 3% $2.50 maximum.		
Arsenic	50¢ per unit over 1%	50¢ per unit over 1%.		
Antimony	$1.00 per unit over 1%.	$1.00 per unit over 1%.		
Bismuth		50¢ per pound over .05%.		

The published tenders or contracts supplied by the smelter to purchase or treat ores and concentrates under stated conditions as to price and other items are called *smelter schedules* (see Table 5.6). The major elements of the schedules are the treatment charges, penalties and payments. The *penalties* are imposed for constituents in the shipment which add to the difficulties and therefore the cost of smelting. Normally, they are graduated according to the content of unwanted material or the deficiency of desired constituents. *Payments* for contained metal values differ materially according to the kind of smelter buying the material and to the nature of the shipment. However, payments are rarely, if ever, based on the full content of a given metal in the shipment, nor on the full market value of the metal at the time of settlement. On the basis of the above facts, it is apparent that, smelter contracts which are subject to specific legal rules of consummation and performance, should be entered into only under the direction of a competent lawyer, and with the advice of metallurgists and engineers who have the full knowledge of the technical matters involved.

Some nonferrous ores are not converted into metal but are sold directly to the ultimate consumers, a good example being the selling of chrome ore to the chemical industry for the manufacture of chromium chemical compounds.

A substantial part of the commerce in nonferrous metals involves scrap which is converted into useful forms of metals. It is estimated that about one-third of all the lead in the United States is used for automobile batteries which have a useful life of only two or three years after which it is scrapped and about 80 percent of the lead is reclaimed and used again.

In the case of industrial minerals such as clays, gem minerals, limestone, phosphate, potash and sulfur, marketing is almost invariably a matter of individual bargaining between producer and buyer. In such cases, the general requirements are indefinite or lacking completely and the working specifications are based on the specific physical and chemical properties of the product. However, in some cases, the requirements are based on the whims of the buyer such as feel, taste, sight, and even sound and smell with or without an additional test of behavior under some manufacturing process.

Because of the uncertainties involved in marketing industrial minerals, a comprehensive market survey is an essential phase of a new industrial mineral project. Such a survey should be made by a reputable firm or a consultant who knows enough about the industry and the technical as well as the economic aspects of the commodity. In many cases, the market survey includes testing of the prospective products by the consumers to prove their suitability. Thus, research, customer services, and product development have become integral parts of marketing of many industrial minerals. This is especially true prior to the introduction of a new product or in attempting to establish a new market for an already established mineral. Sometimes, a cooperative research program, between the producer and the prospective customer, is essential for developing new grades or types of consumer products.

In the case of relatively cheap but large-volume industries such as sand and gravel, *place value* or location of deposits near large consuming centers is of vital importance. However, place value can sometimes be modified by availability of cheap transportation, as is the case for gypsum from Nova Scotia which is brought to consuming centers on the Atlantic seaboard by cheap water transportation. The enormous programs of building public roads, bridges, dams, and other engineering works, have also been responsible for initiating the utilization of available raw mineral resources for producing portland cement in remote areas which could not justify building local cement plants. Thus, availability of markets in particular areas determines the economic feasibility of developing new deposits and building new plants.

Marketing of petroleum and natural gas has some unique characteristics. Since crude oil is not salable in its original condition except to refineries which convert it into a variety of refined products used by ultimate consumers, the entire market for the United States crude production, except for a small volume used as such or exported, is confined to the operating refineries. In general, the industry brings crude oil from a large number of widely scattered production fields to a relatively few strategically located centers where such crude is economically converted into salable refined products. Such refineries are capable of processing a huge volume of crude. To meet this requirement over a period of years, the petroleum industry has provided extensive transportation and storage facilities, along with complex crude oil trading and exchange systems.

In the profitability of refining operations, transportation costs of crude oil as well as refined products play an important part. Therefore, large concentrations of refining facilities are found in coastal areas easily accessible to tanker transport. Such centers are also found in inland areas near populous and industrial areas where the size of the market justifies long distance pipeline connections with producing fields.

The mechanisms involved in the operations of the markets for crude oil include gathering, transportation, and storage facilities; crude oil trading practices; and methods of pricing crude oil in the field. As mentioned earlier, the cheapest form of transportation of crude oil and its refined products is over water routes by tankers. Such vessels transport several hundred thousand tons of cargo each. To illustrate the extent of tanker transportation at present one only has to note that at a single

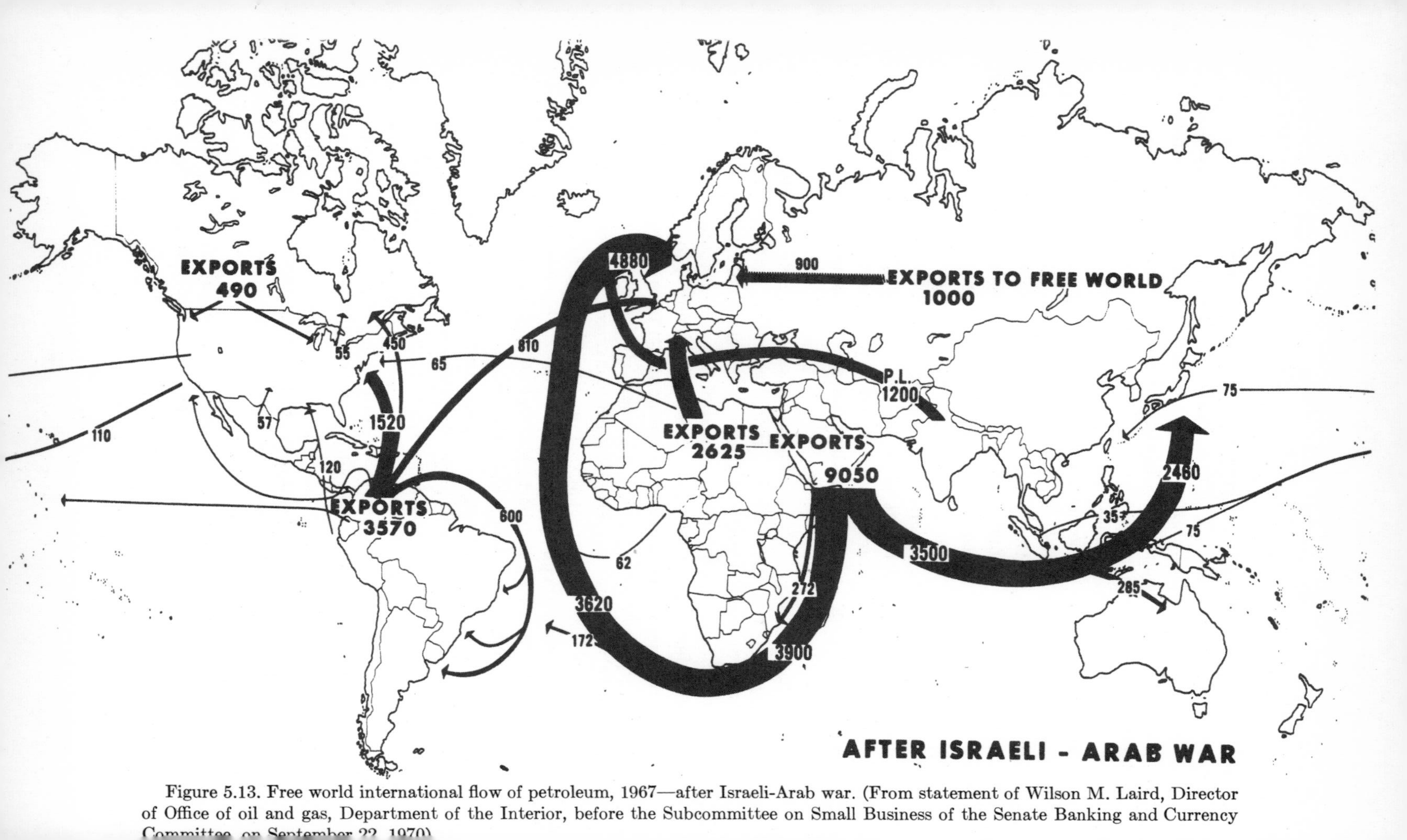

Figure 5.13. Free world international flow of petroleum, 1967—after Israeli-Arab war. (From statement of Wilson M. Laird, Director of Office of oil and gas, Department of the Interior, before the Subcommittee on Small Business of the Senate Banking and Currency Committee, on September 22, 1970)

moment in time about 55 million tons of petroleum are transported over the ocean routes shown in Figure 5.13. In this figure the width of each of the arrows is indicative of the relative amount of oil being transported along a certain route. The Arab-Israeli conflict and closing of the Suez Canal has had an important effect in the transportation practices of the oil industry, and has led to longer routes with the consequent need for larger tankers.

Provision of adequate storage facilities is an essential adjunct to oil transportation by pipeline or over water routes because of the pronounced seasonal variations in the demand for refined products. Increased utilization of heating oil in the winter is a major cause of these variations. Besides permitting accumulation of inventories during low demand, storage facilities allow separation of different types of crude which in turn permits blending of crudes to meet the changing demands of our energy-based society.

One of the common trading practices performed by a refining company owning remotely located crude, is to sell this crude and buy a similar quantity that is better situated with respect to its refineries and market. In such an instance, it is more economical for the company to sell in one area and buy in the other favorable location to save the transportation cost of cross hauling. Such buying and selling activity is called an *exchange.* A refinery, to some extent, is specifically designed for types of crudes that it must process and when a refiner wants a certain kind or blend of crudes for a specialized refinery, trading in crudes enables him to exchange unsuitable oil for a favorable type. It is obvious that all interested parties, the producer, refiner and consumer, benefit from such a trading system.

In general, the price of a commodity or a security is determined through free trading on the floor of an exchange by a large number of buyers and sellers. However, in the petroleum industry, due to the small number of buyers and sellers, oil exchanges have given way to the practice of *posted* prices. Under this system, schedules of crude oil prices specifying location, grades, and gravities of oil are published by buyers. This posted price constitutes an open public price at which crude oil is bought from all producers in a given field, irrespective of the volumes offered for sale. The price posted is a wellhead price and does not include costs of gathering and transportation which is provided by the buyer. Under normal conditions, the buyer takes the entire output of the lease which is fixed as to its upper limits by state conservation laws or by maximum efficient production rates determined by the reservoir engineering studies. Obviously, this is a unique situation in which the price of a commodity is established by the buyer and not by the seller, and the producers have to sell their oil at a price specified by the buyers.

The major factors affecting the pricing of crude oil are:

(1) buyer's prerogative to set the price,
(2) the economic interest of the purchaser which makes him hold the crude oil prices down,
(3) the inability of the seller to hold his production back in an effort to obtain a high price due to the fear of drainage to other neighboring properties,
(4) availability of cheaper crude from previously discovered producing oil fields,
(5) difficulty in determination of producer's current replacement costs, and
(6) adjustment of supply and demand due to conservation regulations.

Concerning the natural gas industry, the rapid growth in the network of natural gas pipelines in the United States after World War II has made this extremely useful commodity available to nearly all parts of the country. As a result, the consumption of gas has continued to increase at a rapid rate until today about 20 percent of the total energy demand in the U.S. is supplied by gas. Currently, there are over 200,000 miles of gas transmission pipelines in operation in the United States and over half the total production moves in interstate commerce. Of the total consumption, about two-thirds of the volume is used for industrial purposes while the remaining one-third is accounted for by residential and commercial uses. From the viewpoint of economics, natural gas has to compete directly with coal and fuel oil in all the available markets. However, inherent advantages of natural gas, such as cleanliness, convenience and high combustion efficiency have made it comparatively cheaper than either oil or coal in many parts of the country.

The producers, the gas transmission companies, and the local distributing companies constitute the three major segments of the natural gas industry. Texas, Louisiana, New Mexico, Oklahoma, and Kansas contain nearly 90 percent of the nation's proven gas reserves and account for about the same percentage of total production. The number of companies engaged in the gas business amount to several thousand and even the largest company owns less than 10 percent of the country's gas reserves. Like oil, the discovery and development of natural gas reserves are a costly and risky business and in general the production of oil and gas is carried on as a joint effort, often from the same field.

Gas transmission companies each year invest a huge sum of money in constructing additional high-pressure pipelines and auxiliary facilities to meet the growing demand for gas. Most of these companies purchase gas from the producers and sell to large industrial consumers and local distributors. In order to make this an economic undertaking, sufficient gas reserves must always be committed to the system to amortize the capital costs and to earn a reasonable profit. For this reason, the usual gas purchase contract is nearly always for a twenty-year term or for the life of the wells. The producers, in order to protect themselves against inflation, have always insisted on an "escalator" clause in the contracts. Such clauses may consist of stated price increases at fixed intervals, price increases to meet higher prices paid to other producers, and adjustments tied to a price index. These interstate pipeline operators are not common carriers but are regulated as public utilities by the Federal Power Commision and the local state regulatory bodies.

In general, local distributing companies buy natural gas from the transmission companies and are guaranteed a fixed area of marketing without competition. They, too, are controlled by state or local regulatory bodies. The gas is delivered to each distributor at a convenient measuring station, where the pressure is reduced to a safe level for consumer use in a given community.

5.5 EXPLORATION OF MINERAL RESOURCES

We have seen that mining is a one crop industry and once the "crop" of minerals is harvested, the operator must find a new deposit elsewhere. It is estimated that in the United States alone, operating mining companies must find at least 110 million tons of new copper ore each year to keep up with production, the figure being three times as much for companies operating in foreign countries. In the case of iron ore, exploration geologists must develop about 800 million tons of new ore each year to satisfy world demands, and it is obvious that these figures will have to be doubled by the end of the century.

The next big question is how will we find these huge quantities of ores? In the past, lone prospectors have been credited with finding many well-known productive mines. However, the days of the individual prospectors are over, since most of the easily encountered outcrops have been discovered and their corresponding mines exploited. The mines of the future are located deep beneath the earth's surface most of which is covered by water, gravel, and sediments. Therefore, there is a need for faster and more penetrating prospecting methods. Also, it is clear that this would be a group effort involving several scientific and engineering disciplines rather than a lone prospector.

In current exploration activities, geologists and geophysicists are using modern airborne equipment for reconnaissance exploration over jungles, deserts, lakes, and seas in a very short period. Moreover, basic research and progress in geological sciences enable an exploration staff to select favorable targets for greater precision. Once the targets are specified, the follow-up ground crews with up-to-date geophysical instruments investigate the area in detail.

Besides using geophysical techniques, the modern prospectors employ several newly developed *geochemical* and *geobotanical* techniques to discover new ore bodies.[36] Such methods measure extremely small quantities of volatile metals (in a few parts per million or even billion) that are emitted from the ore body and form a halo in the air over the deposit as is the case with ore bodies containing mercury. In other cases, minute amounts of some metals migrate from the deposit into streams and even into trees and plants. The last phase in the exploration activity involves diamond drilling to confirm the presence of profitable ore and to outline the extent, shape, and size of the deposit.

The increase in success factor in the search for hidden mineral deposits in recent years can be directly attributed to the dynamic evolution of exploration geophysics as a spin-off resulting from important contributions in the field of space geosciences. The major emphasis today is on the development of super-sensitive detecting, measuring, and recording equipment which allow the pickup of weaker and weaker signals against backgrounds of geologic and instrumental noise.

Generally speaking, all *geophysical* methods are based on measuring dif-

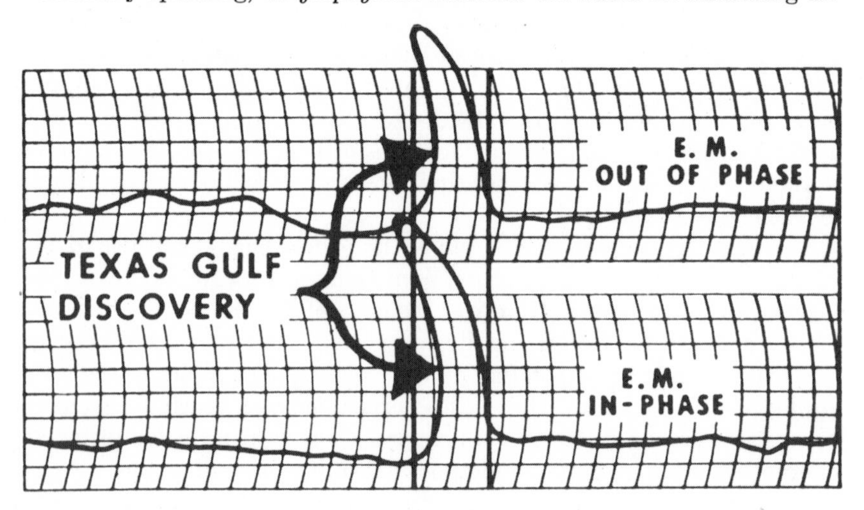

Figure 5.14. Electromagnetic Peaks on recording tapes revealing the presence of a *rich orebody. (By permission of the editor, Engineering and Mining Journal, vol. 167, no. 6, June, 1966.)*

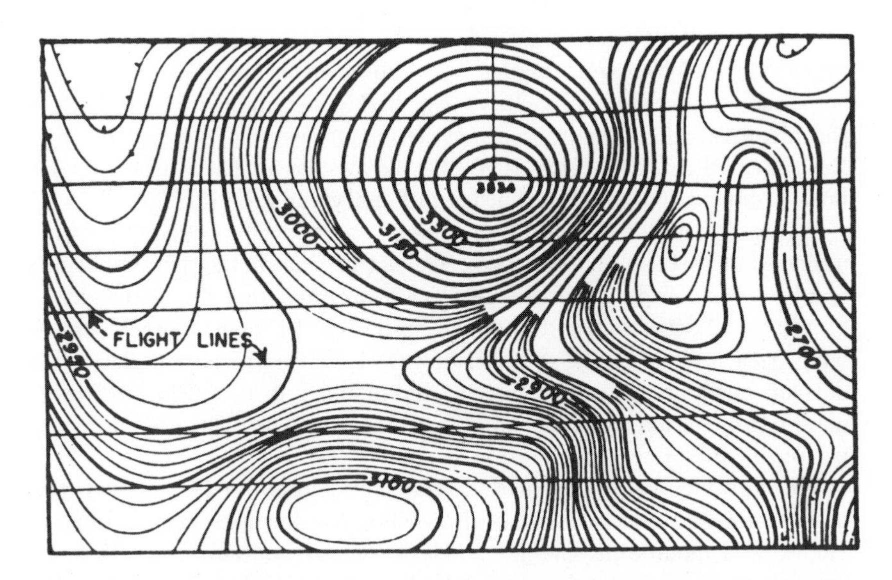

Figure 5.15. Aeromagnetic geophysical survey shows flight lines indicating the path of the plane and gamma contour lines revealing a magnetic high. (*By permission of the editor, Engineering and Mining Journal, vol. 167, no. 6, June, 1966*).

ferences in physical properties followed by interpretation of data into a geologically representative framework. *Electromagnetic systems* are used to measure conductivity of the ground by transmitting a primary electromagnetic field at some low audio frequency and measuring the distortion caused by conductor mineral deposits within the detection range of the system. A recent innovation is the *Induced Pulse Transient System* (*INPUT*) which makes use of a powerful pulse of short duration and a polarity reversal to generate eddy currents in the ground. The decay of these currents is then measured between pulses at six points in time along the decay curve of this secondary electromagnetic field. Such a system is capable of giving 20 to 30 percent greater depth of penetration. Figure 5.14 shows electromagnetic peaks on recording tapes of an exploration helicopter flying over an area north of Timmins, Ontario, Canada. These peaks revealed the presence of a high grade copper-zinc-silver deposit which is currently under development by Texas Gulf Sulphur Co. In this case, the geophysical indication was confirmed by diamond drilling.

Magnetic methods are based on a susceptibility plus remanence factor difference and are useful for both exploration and structural mapping. Instrumentation varies from a small hand-held package to sophisticated airborne systems. The latest development included a unit using the principle of optical pumping of electrons of a gas or vapor and is capable of measuring 0.01 gamma compared with 1 gamma for some of the earlier types of magnetometers. Figure 5.15 shows an example of a successful aeromagnetic geophysical survey. The flight lines indicating the path of the plane carrying an airborne magnetometer revealed a magnetic high. Bethlehem Steel Co., using this clue, discovered a major iron orebody buried under 1,500 ft. of overburden in Pennsylvania.

Gravimetric methods depend on differences in specific gravities between mineralized areas and barren zones. The net increase in mass in the vicinity of a base metal deposit will give rise to a measurable increase in the earth's gravitational attraction in the mineralized area. In the past, such surveys have been limited to the ground; however, efforts are now underway to develop airborne systems. A recent development in this field has been a new high-precision borehole gravity meter which can be used for both petroleum and mineral exploration.

When gravity values are measured at points in an area being surveyed, these values must be corrected to account for the presence of adjacent valleys and mountains, changes in elevation, horizontal changes in mass, changes in topographical configuration of the buried rock surface, horizontal mass discontinuities, and earth rotation. Once all these corrections are applied, a gravimetric map of the area can be constructed and actual anomalies may be properly identified.

Induced polarization (*IP*) methods are dependent on the electrical properties of metallic and electronic conductors embedded in an electrolytic or ionic conducting matrix. When an electric current is passed through a typical porphyry-type copper deposit, the particles of metallic minerals are polarized at their interfaces and this electro-chemical barrier tends to oppose current flow through the ore body, requiring an over-voltage beyond that necessary for a single conductor represented by a barren formation. This over-voltage then constitutes a measurable anomaly indicative of the mineral deposit.

Other electrical methods include *self-potential systems* which measure the electrical energy emanating from decomposing mineral deposits and *resistivity* methods which measure electrical resistance of the area explored. Such electrical survey methods utilize many different instrument designs and electrode configurations for measuring the potential.

Seismic methods are highly developed exploration tools in the petroleum industry employing acoustics for measurement of deep layer effects. However, because of the potential that seismic techniques offer for exploration at depth and over long distances, considerable efforts have been made in recent years to apply these techniques in exploration for mineral deposits. The *audio frequency magnetic* method (Af-Mag) is a newly developed system utilizing electrical disturbances induced by thunderstorms for revealing mineral deposits that are electrically conductive.

In recent years aerial color photographs taken by astronauts have enabled geologists to designate favorable mineral localities based on a perspective study of twisting formations, upended sedimentary rocks, massive intrusions, rock alteration, and discolorations which are all vital clues to mineralized areas. Geophysicists now have new instruments that "see" and "record" images (e.g., infra-red type) and distortions over the entire radiation range. These vary from extremely short gamma rays to the very long radio waves. Furthermore, during the last few years a significant effort has been made to develop geophysical methods and equipment to explore the bottom of the oceans which contain large quantities of minerals and metals.

Realizing the need for the development of more sophisticated exploration techniques and instruments, researchers have intensified their efforts to build direct detection devices such as "*sniffers*" and "*snoopers*" that are specific for tin, silver, beryllium, and mercury.

A serious limitation of geophysical prospecting is the depth of penetration. Efforts are presently underway to extend the range of observations deeper into the earth's crust. The electromagnetic (EM) methods are limited to 100 to 400 feet, while induced polarization (IP) and resistivity surveys to about twice the minimum dimension of the ore body sought. Gravity methods, on the other hand, are sensitive to about 400 feet for mining work and magnetics to about 4 to 5 times the minimum dimension of the ore body. Current and future refinements of instrumentation consist of a variety of systems that are packaged into flying laboratories containing multiple sensors and scanners. However, regardless of the improvements in the sensitivities of the systems employed, the primary reliance in exploration still rests with shrewd geologic interpretation and deduction.

The United States Geological Survey currently uses eleven new search tools and guidelines for exploration. These include:

1. *Neutron activation* which consists of irradiating the surface of the ground with 3 to 14-million electron volt (Mev) neutrons, thus inducing radioactivity in the bombarded area and taking a gamma-ray spectrum for identification of spectra characteristics of contained elements.
2. *Trace analysis methods* for metals.
3. *Mercury halos* which are indicative of sulfide deposits and which may be readily detected and measured by chemical and instrumental techniques (see Figure 5.16).
4. *Remote sensing* from satellite instruments to locate favorable target

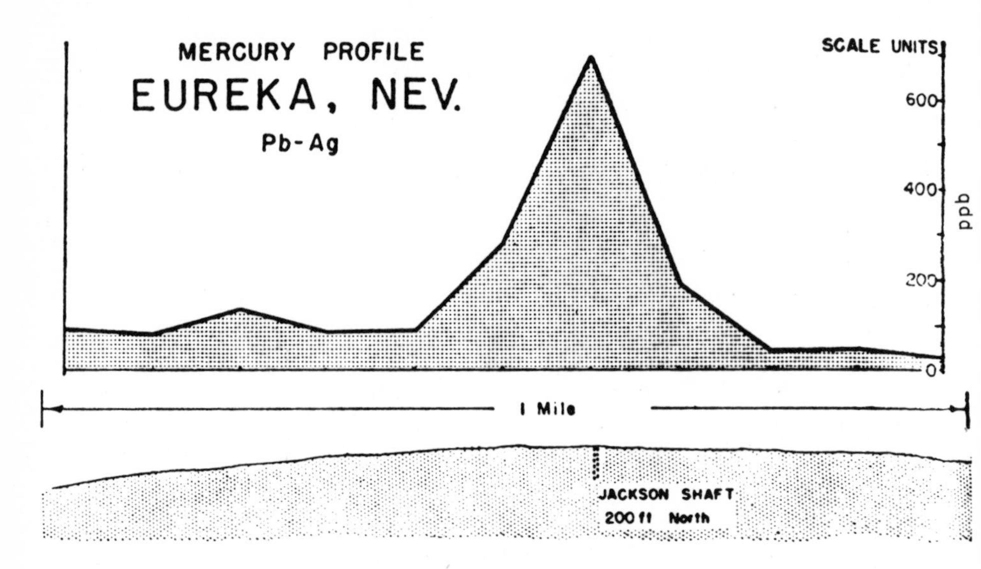

Figure 5.16. Mercury profile indicative of sulfide ore deposit. (*By permission of the editor, Engineering and Mining Journal, vol. 167, no. 6, June, 1966*).

areas through synoptic and panoramic color photography, infra-red, ultra-violet, and radar sensing techniques.

5. Combination of *regional aeromagnetic*, *gravity*, and *gamma surveys* for selecting target areas.
6. *Lead isotope* analysis which provides guides to ore deposits.
7. *Dispersion patterns* of accessory ore minerals which indicate directions of movement of the residual solutions out of an igneous deposit.
8. *Mobile geochemical laboratories* which are totally equipped to perform classical as well as new, rapid, inexpensive analyses for various metals.
9. *Float sampling* as a useful reconnaissance geochemical method in areas which lack surface outcrops.
10. *Scavenged crusts* on stream beds may often provide ore clues due to strong positive correlation between the cold-extractable heavy metal from mineralized areas and manganese (Mn) content of stream sediments from unmineralized terrain.
11. *Gamma-ray spectrometer scanning* helps in locating hydrothermally altered zones associated with copper, lead, and zinc deposits which contain anomalous amounts of uranium, thorium, and potassium. Such scanning devices are mounted in trucks or aircraft for maximum expediency.

In the search for new mineral deposits, exploration personnel have been greatly assisted by the development of a host of new geochemical analytical techniques. Besides previously mentioned neutron activation analysis, these modern methods include: *Atomic absorption spectroscopy* (AAS) which is based on the principle that the atoms of every element absorb light of the same wavelength emitted by that metal when it is chemically unbound and in its minimum energy state. This is accomplished by digesting the sample in suitable media and then vaporizing the metal containing solution in a flame. *X-ray fluorescence* which generates the X-ray characteristics of the sample when it absorbs rays critically shorter in wavelength than the emitted radiation. *Flame emission spectroscopy,* a method similar to atomic absorption in which the element is raised to a high energy state for measurement of radiation. *Direct reading emission spectroscopy* which uses electrical energy to convert a solid or a liquid to the vapor phase in which it will emit light.

Computers, too, have become an increasingly important tool in exploration efforts, especially for interpretation of huge quantities of data obtained through the use of various techniques. They have been effectively used for calculations of ore reserves, for mine evaluation, for modeling data according to given parameters, for predicting trends of structure, for spacing of drill holes and injection wells, and for sorting data. It is quite apparent that the computers' principal advantage for geophysical modeling is in the speed and accuracy of calculation that can be achieved. Flexibility of modeling is another area that the geophysicist can take advantage of in optimizing solutions to their problems. Moreover, computers have important capabilities for recognizing and searching for shapes representing anomalies or signals for known and unknown targets. However, it appears that the most significant contribution that the computers should make in the exploration business in their ability to recognize and evaluate weaker signals relative to noise which will allow the industry to reach greater depths with geophysical systems and perhaps multiply the discovery potential by a factor of two or three over present methods for mineral exploration.

Since the cost of conducting effective exploration programs has been rising steadily, operating companies have shown increasing interest in *operations research,* a modern technique which employs various methods of optimizing the allocation of their financial resources. To this end, mathematical models have been formulated with the intention of quantifying mineral exploration variables in such a manner that they can be integrated will all phases of the company's operations in order to optimize its profit objective. In general, exploration models express at least two concepts one of which is mineral occurrence and the other is effectiveness of search for ore deposits. Harris[37] has proposed a model to predict the unconditional probability of the discovery of "x" deposits in a given area. This model analyzes both the probability to occurrence and that of discovery.

On the basis of the above methodology, exploration personnel are in a better position to inter-relate the concepts of probability, geology, and mineral wealth in each discrete geographic and geologic area. It is apparent that the ultimate use of such modeling is achieved by extrapolation of geologic information from a known area to an unknown one. Thus, by the simultaneous consideration of a number of variables involved in mineral exploration, operational research models can provide probabilities that can help management reduce their error in judgement.

It is also obvious that the recent successes in *space geoscience* will have a tremendous impact on mineral exploration. Adherence to the procedures of lunar-mission-planning involving full utilization of remote sensing techniques is bound to accelerate and to improve detailed scanning of the earth's surface. Lessons learned from the payload capability of the Apollo vehicle system will provide new designs for overcoming the problem of dealing with bulky and heavy electronic equipment now used for geophysical exploration. This should be of direct benefit for field parties operating in wilderness areas of the earth.

5.6 MINING METHODS

Mining may be defined as a process of removing or extracting mineral resources from the crust of the earth. *Surface mining* consists of removing topsoil, rock, and other materials that lie above mineralized zones so that the resource can be extracted. This type of mining offers several distinct advantages when compared to *underground mining* in which the excavation is below the surface of the earth. It provides safer working conditions, allows a more complete recovery of the deposit, and from an economic point of view is considerably cheaper in cost per ton of ore produced (for a typical 10,000 ton/day operation the mining cost for surface mining is 25 cents per ton mined as compared to one dollar per ton for an underground operation). However, surface mining is limited to those deposits in which the cost of removing the thickness of overburden that must be moved in order to recover a given amount of ore is economically favorable.

Surface mining methods employed to extract metal resources include open-pit or open-cut mining, strip mining, glory-hole mining, quarrying, trenching, hydraulic mining, and dredging.[38] *Open-pit mining* is currently used for extracting ores of copper and iron from large ore bodies such as porphyry copper in which the overburden is relatively deep (see figure 5.17).

Figure 5.17. A panoramic view of a large open-pit mine with surface facilities, waste disposal, and water reclamation (*courtesy of Anaconda Corporation*).

In this case, the mining proceeds in a series of contour-like benches as the pit gets deeper and deeper. On the other hand, the term *strip mining* refers to the method used for mining flat tabular beds, such as coal, having a shallow overburden which must be removed prior to mining of coal. *Glory hole* mining is a combination of underground and surface methods. *Quarrying* is the open cut mining method applied to the removal of limestone, building stone, marble, etc. *Trenching* is a form of strip mining used in special cases. *Hydraulic mining* utilizes powerful water jets which disintegrate unconsolidated materials and is extensively used for mining placer deposits containing precious metals. *Dredging* is also a popular method for treating placers and employs a combination of excavation and sorting in a dredge which floats on water. The rejects from the operation are usually put back in water or are piled on the spoilbank area.

Underground mining methods are used for extracting mineral resources which occur at great depths. There are a large number of techniques developed for underground mining. The type of method used depends on the size, shape, depth, type and other characteristics of the ore deposit. In general, underground techniques are broadly classified as caving, stoping, and methods specifically employed in coal mining. In *caving*, the ore is first undercut so that the overlying ore breaks down by its own weight. Sometimes the ore is drilled and blasted to cave the overburden. On the other hand, *stoping* involves drilling and blasting of the ore but overburden is left in place. Under the broad classification of *coal mining* methods one may include the *pillar method*, in which up to 50 percent of the coal is left in the pillars which support the roof, and the *longwall* method, in which the coal is extracted from a continuous face and the roof caves in as the mining advances. *Augering* is also used for coal mining. It involves boring holes 2 to 7 feet in diameter into coal seams that cannot be economically mined by regular stripping methods.

Solution mining has been practiced successfully for quite a long time in the extraction of petroleum and natural gas production. However, it is a relatively new comer in the field of mining. For this latter application, solution mining includes solution of salts, leaching of metallic ores, and the recovery of sulfur by the Frasch process.

The most popular application of solution mining involves the recovery of salt, potash, and nitrates which are commonly mined by dissolving them from their formation by pumping water across the ore deposit. Water is normally injected through strategically placed injection wells and the brine collected from extraction wells.

In recent years, the recovery of metal values from low-grade and submarginal ore dumps by solution mining or leaching has been on the increase. In some instances such as the Kennecott Copper Corporation's

Chino Mines Division Operation at Santa Rita, New Mexico, about 40 percent of the total copper production comes from such dump leaching operations. Another classical example is the Cities Service Company's in-place leaching operation at Miami, Arizona, where leach solution is made to percolate through broken ground in the caved areas.[39] The entire production amounting to about 3 million pounds per month is obtained through in-place leaching operations. In-place leaching of uranium from low-grade dumps and worked-out mines is also becoming popular and in one instance, at the Stanrock Uranium Mines Ltd., Elliot Lake, Ontario, Canada, the entire production amounting to 15,000 pounds per month is realized through such in-situ mining operation.[40] It is quite apparent that with the depletion of higher grade deposits and increasing environmental concern, solution mining techniques may be increasingly used in the future for the production of badly needed metals.

The Frasch process for the recovery of sulfur by solution mining is a well-established method. In this process, hot water is pumped into the ore deposit to melt the element which flows into collection pools from which it is air-lifted to the surface through concentric pipes enclosed in the inlet hot water pipes.

One of the latest mining methods that has gained momentum is ocean mining and recovery of metals from the sea. Because of the great importance of the mineral resources from the ocean, the subject has been discussed separately in the preceding chapter.

5.7 ENVIRONMENTAL EFFECTS OF MINING OPERATIONS

Environmental effects of different methods of mining involve wasteland left by surface mining, pollution of streams by acids and sediment, massive landslides, and deterioration of esthetic values. There is no doubt that some destruction of the environment (air, water, and land) is created through any form of mining, regardless of how efficient the mining method is and in spite of all the necessary precautions taken to abate pollution. Since mining, like any other business, is a profit-making venture, it was, and still is accepted practice to mine as cheaply as possible and this preoccupation with short-term gain too frequently has taken a precedent over the long-term social costs involved, especially in the case of small operators with a lack of capital. This has contributed to the loss of valuable mineral resources and has left derelict large areas of productive land. Today, with the changes in the national sense of values, prodigal waste of our mineral resources or degradation of our environment are not looked at in a tolerant mood.

Because of the inherent nature of surface mining, all such methods produce dramatic changes in the landscape due to open pits, trenches, spoil (overburden) banks, benches, subsidence, and tailing piles from dredging operations[41] (See Figure 5.18). Much effort has been spent by the mining industry in restoring the landscape and in reclaiming the land after surface mining of coal, sand and gravel, phosphate and other materials. Such reclamation techniques, however, are not applicable to large chasms created by open-pit operations. Surface mining also leads to rapid erosion of land and waste dumps because of slope stability problems and denudation of vegetation. Detrimental effects of such erosion are ultimately felt in silting and degradation of streams and waterways. Air and water pollution have also been attributed to dust produced during surface mining through truck haulage, ore handling, and blasting. Two outstanding examples of this particular form of pollution are the phosphate dust fallout in the United States[42] and the lead fallout in Ireland.[43]

A more serious problem attributed to surface mining concerns acid-mine waters as a result of oxidation of exposed pyrite and other sulfide minerals in the mined out areas and dumps[44] (see Section 3.12). Oxidation of these sulfide minerals due to atmospheric conditions as well as bacteria activity converts insoluble sulfide to soluble sulfates and sulfuric acid. Such acid mine waters also contain considerable amounts of dissolved metals such as iron, manganese, and copper which are washed into neighboring streams and rivers. Since the dissolved iron is in the oxidized (Fe^{+++}) state, it precipitates out as undesirable slimy red or yellow ferric hydroxide as soon as the acid is neutralized. Another major problem associated with acid drainage is that even in minute concentrations, soluble salts of zinc, lead, mercury, arsenic, and the like are toxic to fish, wildlife, vegetation, and aquatic insects.

Many streams in the Appalachian region are adversely affected to various degrees by such acid drainage. The Bureau of Sport Fisheries and Wildlife has reported that in the United States some 5,800 miles of streams and 29,000 acres of reservoirs and impoundments are seriously affected by surface mining operations associated with coal mining.

Dredging and hydraulic mining also contribute to water pollution problems which may significantly change the quality of water. Noise pollution, too, has been attributed to surface mining operations, especially those activities close to urban centers. Finally, the economic impact of environmental problems created by surface mining is primarily felt in the reduction of uses and value of properties adjacent to mining operations.

However, it is proper that we acknowledge some of the beneficial effects of surface mining. Of utmost importance is the occurrence of certain desirable hydrologic effects due to the creation of fragmented rock piles

Figure 5.18. Aerial view showing panorama of surface mining for coal. Mined land in background has been planted with grass in reclamation program (*photograph courtesy of National Coal Association*).

during a mining operation. Such artificial barricades prevent danger of floods due to rainfall at the same time allowing the water to sink into the earth to augment ground-water supplies. Some surface mines in the Western United States have exposed ground-water sources which have proved invaluable to wildlife and livestock. In Alaska, gold dredging operations have destroyed the permafrost and the mined areas as well as resulting tailing dumps are considered premium property for industrial and residential developments. Moreover, mine-access roads, when properly maintained have potential for multiple-land-use which benefits large segments of population. Such roads may be used for fire protection, recreation, and management activities, which ultimately increase the value and use of isolated land. Surface mining in many cases has created opportunities to develop recreational areas where none existed before.

The environmental problems associated with underground mining are primarily concerned with subsidence of land due to the caving method used during mining which leaves dangerous and unsightly canyons. The waste dumps created at the mine site as a result of mine development and operation also constitute an additional pollution problem which affects the surface drainage and silting of streams. Underground mining operations along with resulting waste dumps also cause serious acid mine water drainage problems just as for the case of surface mining. One of the most effective methods of preventing such drainage is to seal the mine from the atmosphere upon abandonment to prevent oxidation of sulfide minerals.

Another serious problem associated with underground mining is fire hazard due to oxidation of sulfur minerals which produces intense heat and releases large volumes of toxic gases, fumes, and smoke. Sometimes, coal mine or waste-dump fires burn uncontrolled for years. Therefore, it is important that proper precautions to prevent fires be taken throughout a mining operation.

In general, information on the environmental problems associated with underground mining is much more limited than for surface mining and efforts should be made to determine these so that the necessary environmental considerations could be incorporated in the selection of optimum mining methods for the development of a given mineral resource.

The main environmental problem associated with solution mining concerns the possible pollution of both surface and ground waters by leach solutions of pregnant (metal-bearing) liquors. In typical dump leaching or in-place leaching operations, as practiced in the Southwestern United States, the over-all loss of leach solutions amounts to about 6 to 10 percent.[45] However, much of this loss is attributed to evaporation. Nevertheless, there is a certain appreciable loss of solution and every effort is made to minimize this loss. Land subsidence is also usually involved in solution mining, usually due to the necessity of shattering the ore body by the use of an explosive. In the event a nuclear explosion is to be used for breaking an ore (as has been proposed by the Atomic Energy Commission), radioactive fallout and leach products will be of prime concern.

The recovery of sulfur by the Frasch process also results in land subsidence in addition to the discarding of highly mineralized bleed water into surface waters.

The pollution abatement technology applicable to environmental problems associated with mining operations involves elimination of the problem at its source; reclamation or utilization of the by-products involved and; treatment to minimize the effect of the problem. Typical examples of elimination of the problem at its source include the prevention of burning of coal wastes and deposits as well as development of new mining techniques which will allow effective sealing of mine entrances against the atmosphere to prevent oxidation of sulfide minerals which results in the formation of acid mine waters. Under reclamation and utilization, the by-products recovery of copper from submarginal grade dumps and tailings, and the recovery of fluorine from the phosphate and aluminum industries are classical examples of sincere efforts by the industry to derive maximum returns from our natural resources along with the minimum degradation of the environment. Finally, under treatment to minimize the detrimental effects of mining, the industry should be credited with excellent strides it has made in stabilizing waste and tailings dumps as well as spoil banks to prevent excessive wind and water erosion and to improve their esthetic impact. Such treatment consists of developing a vegetative cover over the exposed surface or stabilizing the surface with synthetic covers.

Recently there has been considerable pressure exerted on the minerals industry to improve their public image regarding pollution and conservation by incorporating sound land reclamation, mine safety and control of air and water pollution. The industry's response to this challenge though not overwhelming, has been very encouraging. Today, with but few exceptions, mining companies are aware of the need to practice environmental control over the waste products stemming from their operations. Since 1918, when the first trees were planted by the Indiana coal industry, remarkable progress has been made in reclamation work.

Basic reclamation techniques such as planting, grading, drainage control, pond stabilization, and the like are standard practices employed by modern mining companies to overcome pollution problems. In such efforts, the primary aims are to eliminate or abate undesirable conditions and to restore land and water resources to more productive uses. In recent years, mined land has been reclaimed for cropland, pastureland, rangeland,

Figure 5.19. Area strip mining with concurrent reclamation. (*courtesy of United States Bureau of Mines*).

wildlife habitat, recreation areas, and for occupancy as residential, commercial, or industrial sites (see Figure 5.19). In the United States alone, over one million acres of valuable land have been rehabilitated in this manner.

Outstanding examples of environmental programs undertaken by the mining industry are: (1) "Operation Green Earth" by Peabody Coal Company[46] which is designed for developing surface mined lands to usefulness and attractiveness as quickly as possible and (2) American Metals Climax Company's program for conservation and pollution control[47], especially the care of landscape at the Urad and Henderson molybdenum properties in the Colorado Rocky Mountains. At the Henderson mine, efforts included careful and expensive placement of plant and tailings dam.

5.8 METALLURGICAL RECOVERY

Metallurgy may be defined as the art and science of extracting metals economically from their ore and other metal-containing products and adapting these metals for human utilization. The science of metallurgy includes all the processes and techniques involved in separating the valuable minerals or metals from the worthless minerals (gangue) as well as the smelting, refining, and working of the recovered metal. It deals with the technical applications of basic laws of physics and chemistry to the concentration, extraction, purification, alloying and working of metals to meet the needs of our industrialized society.

The metal industry today is the second largest enterprise in the world, being exceeded in importance only by agriculture. It is estimated that there are about 6,000 separate metals and alloys currently in commercial use and new alloys are being developed constantly. Today, metallurgy has broken up into a large number of specialized branches, each of which has developed its own voluminous literature. Our intention here is not to provide the reader with details concerning the entire field of metallurgy which is adequately covered in several comprehensive reference books[48–49] but to briefly review the basic processes involved in the extraction of metals from their ores and to look into some of the major environmental problems associated with their recovery.

For our purpose, we shall divide metallurgy into *production* or *extractive* metallurgy and *physical* or *adaptive* metallurgy. The former deals with the concentration of the ore to produce a high-grade product which is subsequently smelted, purified, and refined. In addition, extractive metallurgy is concerned with the recovery of by-products such as gold, silver, selenium or antimony. Physical metallurgy, on the other hand, has to do

TABLE 5.7.

Processes involved in extractive metallurgy

RUN-OF-MINE ORE→MINERAL PROCESSING

Crushing—Grinding
Hand picking—Gravity
Magnetic—Electrostatic
Electronic—Flotation

CONCENTRATES OR UNPROCESSED ORES		TAILINGS (TO WASTE)
PYROMETALLURGY	HYDROMETALLURGY	ELECTROMETALLURGY
Roasting	Leaching	Electrowinning
Calcining	Precipitation	Electrorefining
Smelting	Amalgamation	Fused salt electrolysis
Refining	Cyanidation	
	Ion exchange	
	Solvent extraction	

with the fabrication, shaping, and working of these metals. Table 5.7 lists the various processes involved in extractive metallurgy and shows their interrelations.

The first process to which most of the mined ore is subjected is *concentration*, also called *ore dressing*, mineral dressing, milling, or beneficiation. It is a process of mechanically separating the grains of ore minerals from those of gangue, to obtain an enriched product called *concentrate* containing most of the valuable minerals, and a reject or *tailing* containing the bulk of the worthless minerals. The concentrate is next subjected to a chemical treatment to break up the metal-bearing compounds and liberate the metallic elements. Such chemical processes are quite expensive as a rule, and the cost is proportional to the bulk of material treated. Thus, it is economically more attractive to treat a small amount of enriched concentrate than a large volume of low-grade ore. An essential phase of milling involves reduction of the physical size of discrete particles before they can be collected in separate products. Such liberation is accomplished by *comminution* or *crushing* and *grinding*.

The physical separation of mineral mixtures during concentration is possible because of the differences of certain properties of these minerals. *Hand picking* utilizes differences in color and luster; differences in density allow separation by various *gravity* methods such as tabling, jigging, and spiraling; magnetic separation uses differences in magnetic susceptibility of minerals while *electrostatic* separation is based on differences in conductivity of the minerals; *electronic* separation uses sensitive instrumentation to detect differences in radiation emission, reflectivity, color and luster of minerals; and the differences in surface properties of various minerals are employed in separations made by the *froth-flotation* process (see Figure 5.20).

Once the desirable minerals are concentrated, the next steps in the metal recovery system are (a) to break up the mineral to free the metallic component as a crude metal and (b) to refine this crude product up to commercial grade. This phase of extractive metallurgy is often referred to as *chemical metallurgy* whose three important subdivisions are *pyrometallurgy*, *hydrometallurgy*, and *electrometallurgy*. Figure 5.21 shows a typical flowsheet of the steps involved in the extraction of copper from its ores.

Ever since the first metal was obtained from an ore by the action of wood fire in ancient times, the methods of pyro, or fire metallurgy, have been of significant importance in the extraction of various metals. A large number of metals, such as iron, nickel, tin, copper, zinc, gold, silver, as well as many minor and rarer metals, are won from their ores or concentrates by pyrometallurgical means. The unit processes that use a high-temperature for chemical reaction are: (1) *Calcining* in which heat is utilized to decompose the minerals through loss of certain constituents, as for example, calcination of limestone results in the evolution of carbon dioxide and formation of calcium oxide, an essential step in the manufacturing of cement. (2) *Roasting* in which the concentrates are reacted with a gas, usually oxygen of the atmosphere, at elevated temperature without fusion to effect a chemical change and thereby eliminate some component by volatilization, as for example, copper sulfide concentrates are roasted to produce a sulfate, an oxide or a chloride with emission of sulfur dioxide. (3) *Sintering* on the other hand, refers to a roasting reaction in which oxidation is so vigorous that the heat generated is sufficient to cause incipient fusion of the charge with resulting agglomeration or sintering of particles together. Roasting and sintering steps are usually used as a pretreatment to major extractive processes to follow.

Smelting is a major pyrometallurgical process in which the pretreated material is mixed with suitable fluxing compounds and carbon (as a reducing agent) and reacted in suitable furnaces to produce a worthless slag and a molten crude (impure) metal. The slag, because of its lower specific gravity, floats on top of the molten metal and is readily removed and discarded. This smelting step may be accomplished either in a blast furnace or in a horizontal reverberatory furnace. The final step in metal production involves *refining* in which the crude metal is purified into marketable

Figure 5.20. Photomicrograph of particle-bubble attachment in flotation. Particles of xanthate-conditioned 65/100-mesh galena are clinging to air bubbles and are being rafted to the surface. View is horizontal through the side of a glass flotation cell. 75 X enlargement. (*Photograph courtesy of H. R. Spedden*).

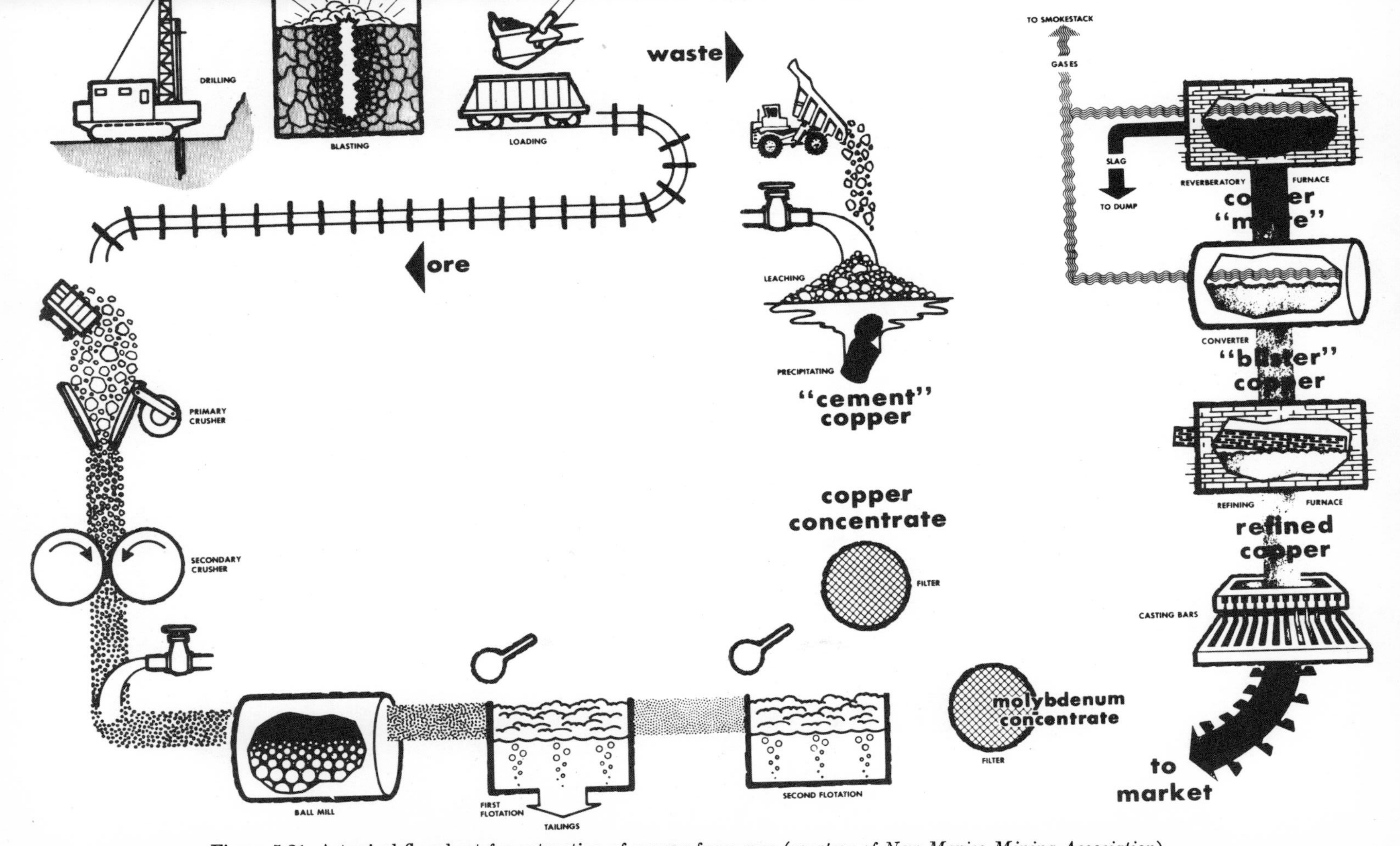

Figure 5.21. A typical flowsheet for extraction of copper from ores (*courtesy of New Mexico Mining Association*).

products. In this process the crude metal is purified either by further heating in the presence of purifying reagents or by *electrorefining*. The finished pure metal is now ready for adapting it for practical uses as a pure metal by itself or as an alloying constituent to produce a large number of commercially usable materials.

For those ores which do not respond favorably to physical separation of valuable minerals from gangue, suitable chemical or *hydrometallurgical* processing is needed to extract a saleable product. In such a process the ores or concentrates are contacted with selective aqueous solvents to dissolve the valuable metals or minerals and then recover them in a pure solid state by appropriate techniques. The basic process followed in this type of metallurgical extraction involves (a) *crushing* and *grinding* of the ore to provide more exposure of the valuable minerals, (b) *roasting* to alter the chemical nature of the mineral so that it will be susceptible to dissolution by the solvent, (c) *leaching* to dissolve the desired mineral by the chosen solvent, (d) *solid-liquid separation* to separate the metal containing solution from the solids consisting of the now-barren gangue minerals, (e) *precipitation or metal recovery* to obtain the metal in a crude or a refined solid state from the leach solution. This may be accomplished either by chemical precipitation such as the cementation of copper by metallic iron, or by *electrolytic* methods.

In the hydrometallurgical treatment of gold and silver ores, specific techniques, such as *amalgamation* involving selective dissolution of precious metals by metallic mercury, and *cyanidation* consisting of the dissolution of precious metals by cyanide solution followed by precipitation of gold and silver with zinc dust, are extensively used. Recent advances in the hydrometallurgical treatment of uranium and copper ores have resulted in an increased use of both *solvent extraction* and *ion exchange*. Both of these processes are dependent on the selective exchange of specific metallic ions from leach solutions onto organic liquids and solids respectively. In this respect these techniques are concentration schemes to recover a particular metallic ion from a low grade leach solution with the exclusion of all major impurities in the process effluent. The adsorbed ions from the organic compounds are then eluted or stripped with appropriate solutions to obtain a high grade liquor devoid of impurities. The purified metals are then recovered from these clean solutions by conventional means. More recently, the versatility of these hydrometallurgical methods has been enhanced by the use of *high pressure* and *high temperature* in suitable *autoclaves*.

The third traditional category of extractive metallurgy is *electrometallurgy* which employs electrolysis to extract or refine metals from aqueous solutions, as in the recovery of electrolytic copper from rich copper sulfate solutions by *electrowinning* or the production of high purity copper by *electrorefining*. In this latter case the crude metal (blister) from the smelter is cast into anodes which are placed in tanks containing an acid electrolyte through which a current is passed. Copper from the anode is dissolved and deposited on the cathode as a pure metal known as electrolytic copper. The residue (sludge) created during this process contains impurities such as gold, silver, selenium, and tellurium which are sometimes recovered as by-products. In recent years *fused salt electrolysis* has become an important technique in the recovery of aluminum (Hall process) and in the recovery of rare earths. The possibilities for future developments of this process are extensive especially if cheap power is readily available.

In looking at the future of metallurgy, it is evident that the new socio-political force will play a dominant role in its development. It will compel us to seek new metallurgical processes that will produce the required metals with a minimum degradation of the environment and maximum conservation of our finite mineral resources. In achieving this goal, Kellogg[50] has suggested three types of improved processes: (1) the bigger and/or better process, (2) the new process by virtue of engineering design, and (3) the new process by virtue of novel chemistry.

The reduction of cost either by increasing the size and output of an existing process or by modification of its performance, has in the past paid handsome dividends. Modernization of the iron blast-furnace over the last three decades, improvement in the Hall process used in manufacturing aluminum, and the computerized electrolytic copper tank-house are classical examples of such innovations. Similarly, new processes by virtue of engineering design should help us in providing improved methods of extraction. Such processes include basic oxygen steelmaking which utilizes a top-blown converter and oxygen gas instead of air,[51] the Imperial Smelting blast furnace for lead and zinc reduction,[52] the cone precipitator for copper cementation,[53] and recent developments in continuous copper-smelting processes.[54]

However, the most important contribution to the future metals technology must be provided by wholly new processes by virtue of novel chemistry. The basic information on the development of such processes can be found in existing technical literature but additional research will be required to put these ideas into practice. Some of the new processes recently developed are; (1) nickel electrowinning developed by International Nickel Company,[55] (2) high temperature-pressure processes for nickel, cobalt and copper developed by Sherritt-Gordon Mines Ltd.,[56] and (3) the solvent-extraction process for copper recovery from leach solution utilized by Ranchers Bluebird Mine in Arizona.[57] There is no doubt that more effective metallurgical processes will be developed in the future based on pyro-, electro-, aqueous-, and vapor-, chemistry as well as on basic principles of physics, geology and biology.

Since the art and science of metallurgy deals in the separation of valuable material from waste products either in the solid, liquid or gaseous form, it is easy to appreciate that mineral processing engineers have a special competence to understand and to remedy pollution situations. The same basic principles that govern metallurgical processes also control pollution abatement technology and most of the common processing machinery can be applicable to pollution problems. Classifiers, cyclones, electrostatic precipitators, thickeners, flotation cells or aerators, digestors, filters, extractors, and the like are all mineral processing units that have found specific applications in the pollution control schemes. For these reasons, it is up to the metallurgists and mineral processing engineers to pave the way for improvements and innovations in the pollution abatement technology and to share their knowledge with environmental engineers in providing the badly needed mineral resources and metals with a minimum degradation of the environment.

5.9 ENVIRONMENTAL EFFECTS OF PROCESSING AND EXTRACTION

The environmental effects associated with processing of minerals are minimal. During crushing, grinding, and screening the major problems are dust and noise. With a proper dust collection system, the air pollution can be kept to a minimum. However, it is very difficult to overcome the problem of noise and even in the most modern mills, noise is a serious offender.

In many concentration or cleaning steps the natural ores are water-washed and screened to remove undesirable materials as it is practiced in coal washing and in the production of sand and gravel. The major environmental problem here is the wash-water discharge which usually contains a substantial amount of fine clays, colloidal material, and dissolved salts. However, the most serious problem associated with concentration involves the disposal of tailings in waste piles or tailing dams.

We have seen in the previous section that concentration selectively removes a small fraction of the total tonnage mined into a high-grade concentrate, thus leaving behind the major portion of the total tonnage in the form of worthless tailings which now have to be discarded. It is clear that the seriousness of the problem increases with decrease in grade since more weight per ton of ore mined has to be disposed of. For example, a plant treating 1% copper ore by the flotation process recovers 60 lbs of concentrates per ton (2000 lbs) of ore processed, thus discarding 1940 lbs in the tailings. For a plant handling 20,000 tons of the above copper ore per day, the total tailings amount to 19,400 tons per day or about 7 million tons per year.

Since under normal conditions, these concentration processes are carried out in aqueous systems, the tailings contain not only the solids but also an equal weight of water and thus result in pollution problems associated with solid waste disposal as well as water pollution problems. The tailings pile or dam ties up a large area of otherwise usable land, is not attractive, constitutes a landslide hazard if poorly planned or engineered, and is a constant source of dust which pollutes the air.

In general, both mining and milling operations leave behind a substantial quantity of solid wastes scattered across the country as unattractive barren piles that create an environmental problem. Realizing the extent of this problem, Congress, in the Solid Waste Disposal Act of 1965 delegated the United States Bureau of Mines to conduct research designed to prevent air and water pollution from such solid waste accumulations. The Bureau has published a large number of excellent reports covering its research efforts[58–60] in this field.

In recent years several waste stabilization methods have been developed and are successfully used in practice. These include physical, chemical as well as vegetative methods. Erosion of waste piles can be prevented by physical spreading of coarse slab, concrete or soil on these waste materials. It is also possible to chemically bond the waste particles together with suitable chemicals such as resins, elastomeric polymers, lignosulfonates, bituminous base, tar, and pitch. However, the vegetative method appears to be the most promising technique for stabilizing waste piles because it is both practical and aesthetically appealing. Several species of plants can germinate, grow, and reseed in mineral-bearing waste material under unduly severe conditions. The modern mining companies have begun to accept land restoration as an integral part of their operations and have utilized the above mentioned techniques with considerable success.

The environmental problems associated with metal extraction are numerous and in some cases quite serious. Under hydrometallurgy, amalgamation (using mercury) and cyanidation (utilizing cyanide solutions), the two major processes for extracting gold and silver, have debilitating effects on the environment. This is especially significant since recent findings of high mercury levels in both marine and fresh water life have caused much concern. However, the extent of this deleterious effect as a direct result of gold mining and extraction has not been fully ascertained. As we have seen in earlier sections (5.3 & 5.5), mercury is geologically associated with heavy metal mineral deposits as halos which aid in the geochemical exploration for minerals and it is entirely possible that much of the mercury found in lakes and the oceans may have been contributed by such natural means. In cyanidation operations, residual cyanides left in the tailings may be washed by rain and surface drainage into receiving streams and made toxic to fish, wildlife, and livestock.

There is also some possibility of environmental situations similar to acid mine drainage arising from copper dump and heap leaching operations which use iron cementation to recover the metal.

Hydrometallurgical processing of uranium ores also poses a serious environmental problem due to residual radon and daughter products emitted from plant tailings. These products are readily washed down with rain and surface drainage and may contaminate streams, rivers, lakes and ground water.

On the other hand, it is understandable that electrometallurgical processes, such as electrowinning and electrorefining, (see section 5.8) are relatively free of environmental problems except for gaseous emissions such as hydrogen, chlorine or fluorine. Such gases are evolved at the anode when aqueous or molten solutions containing the valuable metals are electrolyzed. These gases, however, can be readily captured and reused.

Thus, we find that most of the hydrometallurgical processes do create a water and/or air pollution problem which may be somewhat dangerous to local ecology. However, abatement techniques are available for overcoming these problems either by reclamation, regeneration and re-use of water, or by subjecting the plant effluents to water treatment before discharge. Such treatment procedures consist of lime or limestone neutralization for controlling acidity, oxidation of ferrous to ferric followed by precipitation of ferric hydroxide (chlorine and permanganate treatments),[61] other chemical treatment techniques for complexing heavy metal salts such as coagulation, chelation, and ion exchange, and most-of-all just plain reclamation and re-use of water. It is a well-known fact that in the ore milling and hydrometallurgical operations in the Southwestern United States involving the treatment of copper, uranium, and molybdenum ores, about 85 per cent of the total water utilized in processing is reclaimed and re-used.

It is apparent that the ultimate philosophy in any abatement of pollution should be the by-product recovery of the impurities[62] or the utilization of the waste material involved.[60] It is not sufficient to merely remove the contaminant from plant effluents only to have it pollute the environment once again in another form. It is entirely possible that in most water treatment applications the value of the recovered by-products is in excess of recovery costs. For this reason, recovery of values from mine water, tailings, drainage and process effluents will continue to be investigated since technology and equipment are available for handling large volumes of process waters. Finally, the problems of the past can be solved by available technology, but the solution to problems of the future lies in "environmental evaluation" during the initial planning and development of new ore processing methods.

In pyrometallurgical processing involving high temperature with characteristic emissions of volatile matter, air pollution by particulates and various gases, fumes, vapors, and smoke, constitute the major environmental problems. Dust is particularly a serious problem to cement manufacturers. In the abatement of such dusty conditions physical separation devices, such as centrifugal separators, electrostatic precipitators, bag filters, and wet scrubbers have all proved effective. In *electrostatic precipitators* the fine particles of dust are made to assume an electric charge and are then attracted to an oppositely charged plate. The particles thus collected build up into a layer on the plate and are dislodged at intervals by pounding on the plates. A new unit for control of both gaseous and particulate matter, especially in the submicron range is called the *cross-flow nucleation* scrubber.[63] In this case, the submicron particles are agglomerated into larger particles during the collection process by taking advantage of the excess of free energy of submicron materials. Another innovation along the same line consists of *dry scrubbing* of fine particles which is different in mechanism from wet scrubbing in that the solute gas molecules need to diffuse only through the gaseous phase and not through both gaseous and liquid phases. Thus, the speed of absorption is increased from 100 to 500 times in comparison to that of the wet scrubbing process. The absorbent in this case is made up of a number of inexpensive extended surface solids coated with a selective reagent.

The most serious problem attributed to smelting, however, is the emission of sulfur dioxide from copper, lead and zinc smelting operations and fluorine compounds from phosphate and aluminum processing. These significant air pollution problems still lack in technology for control and abatement. Since the degradation of the environment by sulfur dioxide emitted from smelter stacks has received much attention in the literature,[64–65] it would be appropriate to look into this problem at some length.

According to Swan,[66] the annual emission of sulfur in the world is estimated at about 212 million tons of which 28 percent or about 60 million tons is derived from man-made sources. The smelting operations contribute about three percent of the total sulfur (6.5 million tons); the emission in the United States being about 2 million tons in 1970.

If we examine the sulfur cycle we find that it is a continuous one in which

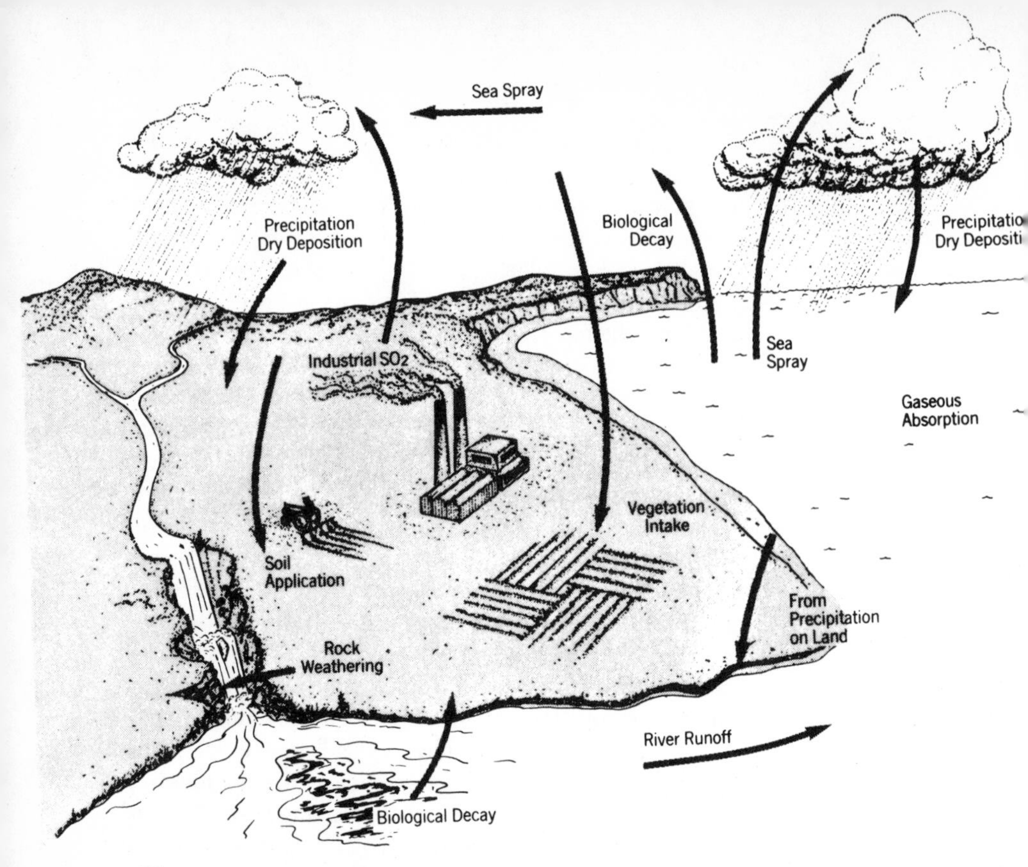

Figure 5.22. The sulfur cycle (*By permission of the editor, Engineering and Mining Journal, vol. 172, no. 7, July, 1971*).

sulfur emitted to the air undergoes transport, deposition, absorption, precipitation and assimilation by biological processes before returning to the land and the sea (see Figure 5.22). The life of sulfur compounds in the atmosphere is known to vary from four hours to four days depending on climatic conditions.[67] For this reason, it is understandable that there is no evidence of any accumulation of SO_2 in the atmosphere and that very little of the contaminant escapes from the immediate vicinity of the smelter.

The Environmental Protection Agency has proposed the National Air Quality Standards for six common classes of air pollutants: sulfur oxides, particulate matter, carbon monoxide, photochemical oxidants, nitrogen oxides and hydrocarbons. These standards are published in the *Federal Register*, January 30, 1971, and will apply to all areas of the United States. The proposal covers both the national primary standard designed to protect human health, and the secondary standards designed to protect against effects on soil, water, vegetation, materials, animals, weather, visibility, and personal comfort and well-being. The proposed National

Ambient Air Quality Standards for SO_2 are as follows:

Standards	SO_2, Annual		SO_2, 24-hour	
	mg/M^3	ppm	mg/M^3	ppm
Primary	80	0.031	365	0.139
Secondary	60	0.023	260	0.099

The value expressed in micrograms per cubic meter cannot be exceeded more than once in any 24-hour period during the year. Besides establishing the air quality standards, the Clean Air Act Amendments of 1970 also specified a timetable under which these standards must be achieved over a period of five years.

Can the smelting industry achieve these standards in the specified period? The answer to this question is not a simple one. The waste gases emitted from a typical copper smelter can be classified into two general categories, namely, a relatively weak gas of 0.5 to 3.5 percent SO_2 content at a high volume from the reverberatory furnace and a 5 to 6 percent gas at a lower volume from the converting operation. The objective of an effective emission control system in smelting, is to convert the SO_2 into a useful or disposable product such as sulfuric acid or elemental sulfur. Production of acid is feasible only if a local outlet for it is available. If not, elemental sulfur is the preferred product since it can be shipped to a distant market or can be stored if it cannot be disposed of readily.

Theoretically, sulfuric acid can be manufactured from either the weak or the strong sulfur-dioxide-containing smelter gases. However, the physical size of needed facilities for producing the acid become impractical with weaker SO_2 streams. The major drawback here is the lack of consumers for the acid within practical access of the smelters. It is estimated that about 6 million tons of acid would be produced if all the smelters in the southwestern United States converted their gases to the acid form. Currently, the demand for acid is only about a million tons, and therefore, such a large portion of the corrosive product cannot be disposed of while its storage would be totally impractical. On the other hand, several proven processes are available for producing elemental sulfur from smelter gases containing more than 18 percent SO_2. Attempts to produce the same product from lower gas concentrations have not been successful in spite of a tremendous effort over a number of years.

According to an in-depth study of the SO_2 problem by Fluor Utah Corporation,[68] the currently available proven technology can achieve ambient air quality standards, but cannot meet the 90 percent emission control standard which specifies that the maximum allowable emission be based upon 10 percent of the total sulfur feed. It was also suggested that the conversion of smelter gases to sulfuric acid would require an industry-wide investment of $607 million which in turn will raise the production cost of copper by 5.2¢ per pound. Finally, the study indicated that both ambient and emissions standards could be achieved with a combination of the sulfur oxides recovery systems with caustic scrubbers in series with limestone scrubbers. However, this system is still in its experimental stage and will require more than three years to develop.

The problem of prevention of SO_2 emissions from the smelters has long been of concern to the industry, government agencies, and research organizations. During the last few years, millions of dollars have been spent in research and over one hundred different processes for the recovery of sulfuric acid or elemental sulfur have been proposed and tested. These include improvements in smelting techniques, such as flash,[69] electric, blast, and continuous smelting as well as non-smelting hydrometallurgical processes, such as high temperature-pressure leaching,[70] dissolution by ferric chloride,[71] Treadwell process,[72] bacteria leaching,[73] and the like. Recently, the eight major copper companies have joined their resources to form the Smelter Control Research Association whose primary objective is the development of improved methods of recovering sulfur dioxide and particulates from smelter stacks. This program should enable the mining industry to find suitable solutions to the SO_2 problem in the shortest possible time. It appears that operational systems will be available to the industry in five to seven years. This case sets an example as to how government agencies and mineral companies can exercise greater understanding and better cooperation so that the general public will reap the benefits of increased metal production with a stable ecology in a reasonable time.

5.10 ENVIRONMENTAL IMPACT OF OIL AND GAS DEVELOPMENT

In regard to the environmental problems associated with the development of our energy resources, we have already looked at the major problems encountered in coal production under mining (5.7) and processing (5.9). Let us now briefly review some of the major effects that oil and gas production have on our environment.

Specific environmental problems which have been attributed to oil and gas development are:

(a) poorly designed roads and pipelines,

(b) geomorphological or aesthetic damage,
(c) water pollution, vegetation kills, and damage to fish from oil spills and briny waters, and
(d) air pollution.

The oil industry in general has devoted considerable manpower and money towards the prevention of *oil spills* and abatement of accidental oil spills on the water and the land, since the well publicized Santa Barbara incident in January, 1969.[74] This effort has resulted in the development of new techniques and increase in inventory of cleanup equipment as well as in the technology of control and clean-up operations. The establishment of Clean Seas Incorporated by major oil operators in the California area indicates the sincerity of the oil companies to protect the environment around their operating localities and to improve their existing capability to fight massive offshore oil spills.

Such oil spills constitute a hazard to the marine habitat since oil has a tendency to adhere to the gills of fish and interferes with respiration. Oily products and emissions are also very toxic to algae and other plankton thereby destroying benthal organisms and spawning grounds.[75]

Another problem associated with oil and gas production is air pollution which is caused by well blowouts, volatilization of oil spills and the emission of foul gases from oil refineries. However, oil companies in cooperation with the Environmental Protection Agency and State authorities have exerted considerable effort in controlling and abating air pollution.

The oil and gas industry in general has looked at these environmental problems with concern and interest and have done their best to abate the detrimental effects with sound environmental policies and effective research and development programs to solve the problems. The Federal Government, too, has provided safeguards and protection through the National Environmental Policy Act (NEPA) of 1970,[76] Water Quality Improvement Act of 1970,[77] the Refuse Act of 1971,[78] the Clean Air Act Amendments of 1970,[79] and regulations concerning Spills of Oil and Hazardous Substances under the Water Quality Improvement Act of 1970.[80]

5.11 NEW CHALLENGES IN MINERAL EXTRACTION

In addition to recovery of metal values from scrap[81–86] which is discussed in the following chapter, biological and chemical mining seem to be the most promising new techniques for mineral extraction.

Biological Mining

The development of extraction techniques for low-grade ores and rare metals is of paramount importance today and will, no doubt, become even more significant in the future as our high-grade ore deposits are depleted.

In their effort to develop faster, cheaper, and more efficient methods for extracting valuable metals from ores and other metal-bearing materials, mineral engineers have utilized many of the techniques of physics, chemistry, and chemical engineering, as well as of metallurgy.

In recent years, mineral processors have turned to still another scientific area, the field of bacteriology. Lately, microorganisms appear to have gone into business. Imperceptibly, their myriads have crept into various industries, where they are nourished and pampered for the work that they do. Increasing use of the effects of certain bacterial growths is being made in industry, and "germs" are being regarded more and more as the servants of man. Every living cell may be likened to a tiny chemical factory that forms, by means of its various enzymic and metabolic activities, varied and often extremely valuable substances or changes.

Geological studies[87] have indicated that microorganisms are a factor in the formation of certain ore bodies, such as bog-iron and manganese deposits. The native sulfur found in the caprock of some salt domes very likely derived from bacterial conversion of the sulfate content of anhydrite to free sulfur. It is also recognized that the anaerobic type species *Sporovibria desulfuricans* reduces sulfate to free sulfur or sulfide.

In the open-pit mining operations at Bingham Canyon, Utah; Miami, Arizona; Santa Rita, New Mexico; Rio Tinto, Spain; and other similar localities throughout the world, the recovery of copper by leaching waste rock with water has long been a profitable operation. In these operations, wastewater from the copper mines has been percolated and repercolated through waste dumps, and even back through the worked out areas of the mine, to leach out the copper from this low-grade material. Copper from these solutions has been recovered by scrap iron precipitation. In some cases, very low-grade ore deposits have been leached in-place in this manner. Generally speaking, however, this procedure has been regarded as far too slow to be economically efficient if any other method of recovery is applicable.

Ferric sulfate and sulfuric acid naturally present in the mine water, as well as the sulfuric acid produced by atmospheric oxidation of the sulfuric material, are responsible for the solution of copper. Ferric sulfate is also an effective lixiviant for a number of other metallic products.

Following the roast reduction of iron-bearing chromite and titanium materials, metallic iron can be removed by ferric sulfate leaching. Although lead sulfate is not readily soluble in brine solution alone, its solubility is enhanced greatly if ferric sulfate is present in the brine. Ferric ion in solution has proved to be valuable in the purification of electrolytic zinc and manganese solutions, the ferric ion serving as a carrier by removing impurities from the solution.

In all the above uses, the ferric ion is changed to the ferrous state by the reactions taking place during the leaching procedure.

The principal deterrent to any widespread use of ferric sulfate-sulfuric acid leaching has been the need to regenerate the spent lixiviant before recycling it for reuse. In acid solutions, regeneration poses a considerable problem. Atmospheric oxidation of acid ferrous solutions to the ferric state is slow and not aided materially by aeration. Although regeneration has been accomplished by the use of SO_2 and aeration, the use of ferric sulfate-sulfuric acid leaching has been greatly hampered and limited by the need to provide the required SO_2.

In 1947, Colmer and Hinkle[88] were the first to recognize the role of bacteria in the oxidation of ferrous ions as well as metallic sulfides. Their study concerned acid mine drainage from West Virginia's coal mines. Similar studies[89] on drainage water from coal mines in the bituminous areas of Pennsylvania have proved the existence of iron-oxidizing autotrophic bacteria that oxidize the ferrous ion to ferric much faster than can be done by atmospheric aeration.

In recent years, laboratory studies at Brigham Young University[90] have suggested that the production of soluble iron, copper and sulfuric acid in the dump leaching of copper ores was perhaps not due to a simple chemical process but to the action of microorganisms. Similar investigations at the University of British Columbia[91] have confirmed this view and have demonstrated that the production of acid and of soluble iron is indeed a microbial process.

More recently, research by Kennecott Copper Corporation on acid mine waters at Bingham Canyon and Santa Rita has proved the existence of similar strains of iron-oxidizing bacteria.[92] The U.S. Bureau of Mines has been active in biological mining and its research program[93] has resulted in the development of four strains of bacteria that will specifically leach manganese.

The microorganisms involved in biological mining are *chemolithotrophic* bacteria which do not use organic matter but derive their energy from the oxidation of inorganic salts such as ferrous iron, sulfur, sulfide and thiosulfate. In addition, they require carbon dioxide as the sole source of carbon for synthesis of cell structure and oxygen (aerobic conditions) to carry out their life processes. Such bacterium are now generally classified as *Thiobacillus ferrooxidans* and they are found universally both in natural leaching situations and those created by man. As shown in Figure 5.23, they

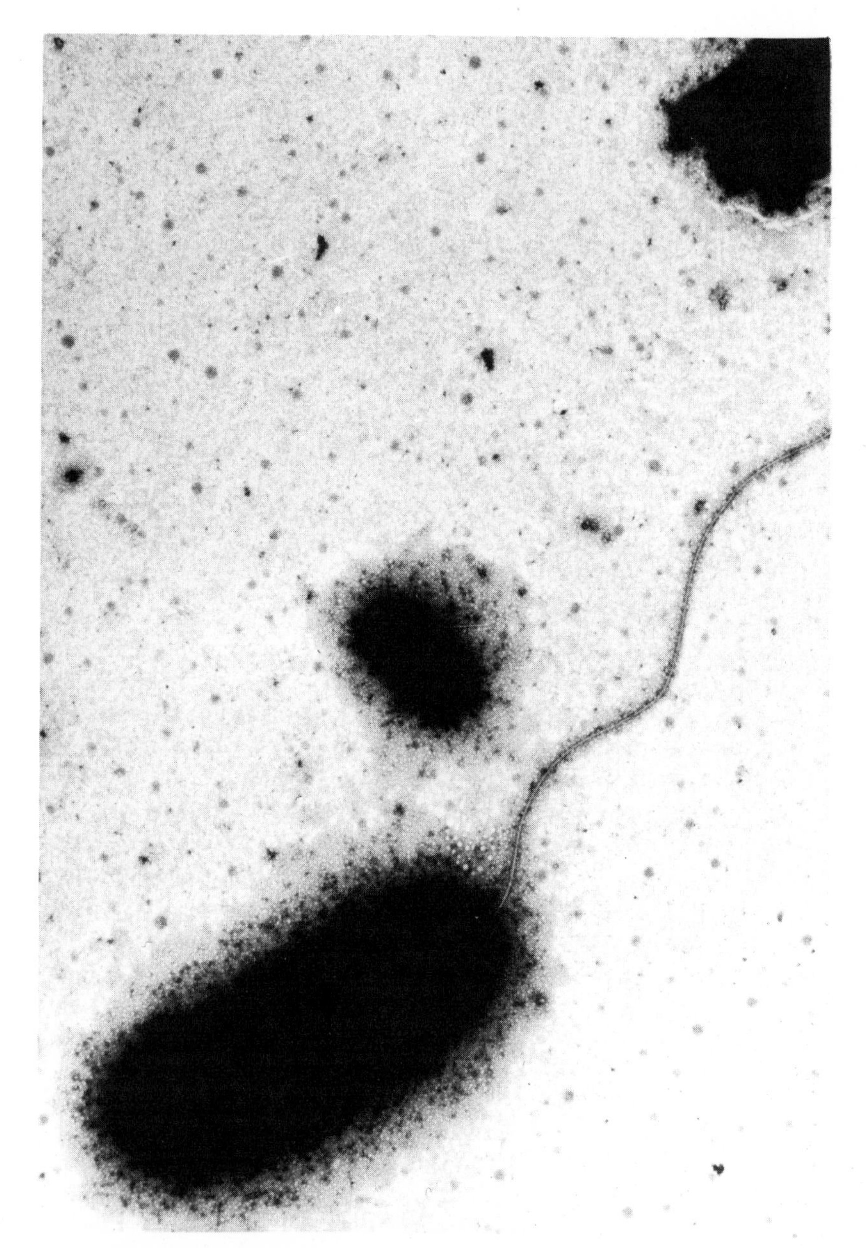

Figure 5.23. Electron micrograph of rod shaped thiobacillus species, 250 X enlargement (*courtesy of K. L. Temple and J. A. Brierley*).

derive their energy from the oxidation of ferrous iron to ferric according to the equation:

$$4FeSO_4 + 2H_2SO_4 + O_2 \xrightarrow{\text{bacteria}} 2Fe_2(SO_4)_3 + 2H_2O \quad (5.11\text{-}1)$$

Since the end product of this reaction is ferric sulfate, a selective solvent for sulfide minerals, these bacteria produce the required lixiviant as a result of their oxidative energy metabolism.

In cases where the acid is entirely used up by the leaching process without a corresponding decrease in iron content, the ferric sulfate produced above may hydrolyze, precipitating the ferric salt and liberating more acid. Thus,

$$Fe_2(SO_4)_3 + 6H_2O \rightarrow 2Fe(OH)_3 + 3H_2SO_4 \quad (5.11\text{-}2)$$

Such hydrolysis can be effective if there is enough iron present. Where sufficient iron is not available, it may be necessary to add sulfuric acid to protect the iron make-up of the solution. Iron may be added in the form of pyrite.

An additional strain of bacteria found in the copper leaching operations belongs to a group called *Thiobacillus thiooxidans*. Although the role of such bacterial species in leaching operations is not well understood, their importance can probably be attributed to their ability to produce sulfuric acid from the oxidation of reduced sulfur compounds according to the following reaction:

$$2S + 3O_2 + 2H_2O \xrightarrow{\text{bacteria}} 2H_2SO_4 \quad (5.11\text{-}3)$$

The sulfuric acid thus produced can be utilized for dissolving oxide-copper minerals and for maintaining proper acidity for the leaching process.

For optimum activity these thiobacilli require an acid medium. It is possible to enhance their bacterial action by adding certain nutrients to the lixiviant. Nitrogen in the form of nitrate or ammonium ions increases the initial rate of activity but apparently does not affect the later stages of the reaction. Recent studies at New Mexico Institute of Mining and Technology[94] have shown that certain organic compounds containing ammonium and phosphate ions do increase the bacterial activity appreciably.

The activity of the bacteria is also affected by light and is increased by operating in total darkness. Usually bacterial activity is decreased by lowering the temperature. For this reason, temperature must be controlled to avoid killing or immobilizing the bacteria. A maximum practical operating temperature of 40°C has been suggested.

The optimum pH appears to be 2.0, but excellent results have been obtained with a pH range of 1.5 to 2.5. Bacterial activity decreases progressively and significantly below a pH of 1.5.

It has been found that the bacteria strains naturally occurring in mine water can endure and grow in a relatively high concentration of dissolved copper ion, though having low tolerances for certain other metal ions. For example, naturally occurring *Thiobacillus ferrooxidans* will not grow in solutions containing more than 150 ppm zinc, even though the species will survive and thrive in solutions containing a higher concentration of copper.

By breeding successive generations of these bacteria in culture media having higher concentrations of other dissolved metals, it is possible to produce new strains that will tolerate solutions containing relatively large concentrations of such metals. For example, bacteria have been developed that have a zinc tolerance as high as 17 grams per liter (17,000 ppm, compared with only 150 ppm in the parent strain). Moreover, the tolerance to copper has been increased to about 12,000 ppm. It is likely that these tolerances can be increased even more.

Particular strains of bacteria have been developed having the following metal tolerances: aluminum, 6,290 ppm; calcium, 4,975 ppm; magnesium, 2,400 ppm; manganese, 3,280 ppm; and molybdenum, 160 ppm.

In the case of bacterial leaching of manganese ore,[93] the U.S. Bureau of Mines found it necessary to develop strains that would act directly on the manganese minerals, because of the absence of sulfide minerals. Their work indicated that only certain cultures were able to effectively leach manganese. With this method it was possible to recover 97.5 percent of the manganese contained in ores averaging 3 to 4 percent manganese in a 60-day leaching period.

It should be noted that for a particular species to perform as a leaching agent for manganese, it may be necessary to obtain a strain or subspecies that has been conditioned to a manganese environment. Several years of conditioning may be required.

One of the latest applications of biological mining involves the extraction of uranium from low-grade ores in the Elliot Lake area, Ontario,[95] as well as in other parts of the world. In all such cases, the bacterium involved in the leaching process has been determined to be *Thiobacillus thiooxidans*, a strain capable of oxidizing ferrous iron and metal sulfides. The role of bacteria in this oxidation process appears to be limited to the regeneration of chemical oxidant as well as ferric iron and the production of sulfuric acid which is an effective solvent for uranium minerals. Because of similar mechanisms, the bacteria leaching of uranium can be optimized under the same conditions as are needed for copper leaching.

From an environmental standpoint, biological mining appears to be an attractive extraction technique. First, the mining of ore from its original location is avoided thus eliminating the problem of degradation of land as well as disposal of waste rock and tailings. And, secondly, such a process does not pollute air or water. Of course, there is always the danger of contamination of groundwater during bacteria leaching. However, because of reliable control over properly engineered operating conditions, this problem is avoided or kept to a minimum allowable level.

Because of its advantages, bacteria leaching has been investigated as an alternate to smelting of copper sulfide concentrates. In this procedure, the sulfide minerals are fermented in a matter of hours, rather than in days or months without creating smoke and related environmental problems. This method has been successfully applied to pilot-scale fermentation of sulfide concentrates in large tanks and the chances for successful development of a commercial process in the future appear to be favorable.

Chemical Mining

Chemical mining is the *in situ* extraction of metals from ores located within the confines of mines, dumps, prepared ore heaps, slag piles, and tailing ponds on the surface. These materials represent an untapped, potential source of all types of metals. It is not inconceivable that eventually our ore reserves will consist of largely low-grade, refractory, and inaccessible new deposits and low-grade zones near previously worked deposits, caved and gobfilled stopes, waste dumps, tailing ponds, and slag piles. Chemical mining promises economic recovery from these types of sources.

Up to now, this kind of mining has been more or less limited to the extraction of copper from low-grade materials. It probably has a much greater potential than this since most metals are susceptible to leaching in the *in situ* environment.

Any process used in mining or mineral processing has certain advantages and disadvantages. A few for chemical mining are listed below:

Advantages

1. Chemical mining can be often used to recover metals economically from materials that could not be so treated by more conventional mining, milling, and smelting processes.
2. A chemical mining operation usually requires less capital investment than a mine and mill.
3. A chemical mining process usually increases a mine's ore reserves. Low-grade or inaccessible ore zones, gob and caved fill, and dumps and tailing may become ores.
4. The leach liquors obtained through chemical mining usually lend

themselves to a variety of metal recovery processes. The pure metal or metal compounds so obtained may be of greater value to a mine owner than the sulfide or oxide products normally obtained by conventional milling processes.

5. Chemical mining may prove applicable to recovering metals from ores that are too refractory for conventional processes.
6. Chemical mining can often be used in conjunction with a conventional mining or milling process to boost metal recoveries and increase ore reserves.
7. From an environmental standpoint, chemical mining appears to be an attractive alternative to currently used mining methods which create serious pollution problems.

Disadvantages

1. Both physical and chemical restrictions may limit the usefulness of a chemical mining process. The effectiveness of contacting ore with solutions and recovery of leach solutions from the system without appreciable loss are two important physical factors. Dissolution or dissolution rates, metal precipitation, and solution regeneration chemistry are major chemical factors.
2. Testing a chemical mining process short of actual field operation sometimes proves difficult.
3. Ground-water contamination may result from some chemical mining operations.
4. We presently lack basic information on the physical and chemical factors involved in chemical mining.

The field of chemical mining can be divided into (1) mining economics and ore evaluation, (2) elements of the leaching phase, (3) preparation of ores, (4) practical aspects of *in situ* leaching, (5) reagent generation and regeneration, and (6) recovery of metals from leach liquors.[96]

In considering the economic exploitation of a deposit through chemical mining, one must determine the size of the deposit, tonnage of ore in place, and amount of metal contained therein. In past as well as present mining operations, the cut-off grade has been governed by the total operating cost, including mining, which usually constitutes a significant portion of the over-all cost. However, in chemical mining considerations, the cost of mining would be minor and, therefore, the cut-off grade would be lowered correspondingly. This lowering would inevitably increase the tonnage as well as the metal content of the deposit, which in turn would influence the over-all economics of the venture.

Lasky[97] and others have studied this relationship through a statistical analysis of known deposits and perusal of past records of some mining companies. These studies reveal that there is an exponential relationship between grade and tonnage of ore reserves. Especially for deposits in which there is a gradation from relatively rich to relatively lean material, there appears to be a consistent mathematical relation between tonnage and grade, according to the equation

$$G = K_1 - K_2 \log T \tag{5.11-4}$$

where T is the tonnage produced at a given time plus the estimated reserves, G is the weighted average grade of this tonnage, and K_1 and K_2 are constants to be determined for each deposit. Using equation (5.11-4) it can be shown that for a typical porphyry copper deposit, the tonnage increases at a compound rate of 14.9 per cent for each 0.1 per cent decrease in grade. Such a relationship has been extended to other metals such as molybdenum and uranium.[96]

These data indicate that the known reserves and metal contents may be increased significantly through corresponding lowering of the grades of the deposits. Chemical mining schemes allow such lowering of grade while increasing tonnage and metal content.

Another important aspect of chemical mining that needs further scrutiny is determining a minimum reserve and grade of a deposit for profitable exploitation. Hardly any data are available on this subject. The only guideline worthy of consideration is the operational data from copper dump leaching and in-place leaching practices in which the grade of material treated is above 0.16 per cent.

The important factor in chemical mining, as in dump leaching, is making sure that the major portion (+90%) of the specified volume of leach solution fed to the deposit or dump is recovered with a given minimum amount of metal content in solution over the life of the economic operation. This minimum metal content in the specified volume is such that the value of the recovered metal will provide for the cost of operation, amortization, and profit.

Naturally, the metal content in leach solution differs from metal to metal, depending on its price. From a hydrometallurgical recovery viewpoint alone, it is estimated that at the current prices of metals and operating conditions, the break-even contents for a minimum operation of 200,000 gallons a day are 250 ppm (0.25 g/1) copper, 50 ppm (0.05 g/1) molybdenum, and 10 ppm (0.01 g/1) uranium. If the mining and development cost, overhead, and profit amount to 200% of the metallurgical treatment cost, then the metal contents in leach solution must be 750 ppm (0.75 g/1) copper, 150 ppm (0.15 g/1) molybdenum, and 30 ppm (0.03 g/1) uranium for an economic operation.

In general, it may be safe to assume that because of lower treatment and capital costs incurred in chemical mining, one could economically treat a sufficiently large deposit containing half the grade of deposits currently mined and milled. Thus, deposits containing 0.25% copper, 0.12% molybdenum, and 0.1% uraniuin could profitably be mined with this technique. In actual practice, it may well be possible to treat even lower grade deposits than these.

Accessibility, physicochemical interaction, and transport constitute the elements of the leaching phase involved in chemical mining.[98] Limitations imposed on any of these factors restrict the leaching process.

Prior to chemical mining, fragmentation of the ore may be needed in order to permit the flow of the leaching solution. In recent years, the Lawrence Radiation Laboratory[99] has proposed the use of nuclear explosives. Griswold[100] suggested using hydrofracting techniques to break ore for subsequent leaching. As envisioned by him, liquid explosives would be injected into ore bodies along planes of weakness and detonated at a slow rate. Although conventional mining methods have seldom been used to prepare ore for *in situ* leaching, there is no reason they could not be used.

Varying degrees of preparation may be required when the ore is already broken. When it is located underground as stope fill, very little ore preparation is required. Waste dumps, slag piles and the like may require crushing and stacking on prepared pads, whereas fine materials like tailings may require rebedding, slime removal, or placement on an impervious pad.

In chemical mining, large volumes of ore contact relatively large volumes of leaching solution over a period of time. The mechanics of solution flow varies according to the reactions involved, the means available for solution containment and recovery, and the need to prevent groundwater contamination.

Two principal types of solution-flow through a porous ore bed are downward percolation under the influence of gravity and flow within an immersed system.

In a downward percolation system, leach liquor is usually distributed over the top of a pile of ore and allowed to flow through the pile to a liquor collection system. This type of solution-ore contact has the advantage that only the floor of the ore bed needs to be impervious to the leach solutions. This type of flow allows some circulation of air within the ore bed, possibly an important factor in the oxidation of ore minerals. Use of downward percolation in an underground ore bed could prevent solution seepage into the groundwater strata. The chief disadvantage of such a system is that incomplete solution-ore contact can result from the impermeability of local zones and from channeling. An outstanding example of this type of in-place leaching is the underground leaching operation of Cities Service Company's copper mine at Miami, Arizona.[39]

Immersion techniques have been used in chemical mining. Utah Construction Company[101] recently conducted leaching tests on an unmined uranium ore body by using a series of injection, monitor, and recovery wells to force a leaching solvent through the permeable uranium ore body and to recover the pregnant leach solutions. Pirson[102] has proposed a similar technique for the *in situ* leaching of phosphate beds while Davis and Shock[103] have investigated a unique technique for solution mining of thin-bedded potash.

Certain features inherent to an underground environment can aid leaching. One is the hydrostatic pressure imposed on an immersed deposit at depth and another is the natural increase in rock temperatures with depth. If gas(es) were introduced into an inflowing stream of solution entering a deposit under a hydrostatic pressure and/or the rock temperatures were above the normal temperatures at the surface, one could possibly use this system as a huge, high temperature-high pressure autoclave. Leach reaction rates can usually be increased many times when temperatures and gas pressures are increased.

Figure 5.24 diagrammatically illustrates how autoclave conditions might be imposed on an immersed mine. This system consists of an abandoned mine flooded with a leaching solution. A pipe introduced into the bottom of the mine via a shaft would carry a gas (usually air) and/or spent leach liquor into the bottom of the system. These conditions would greatly enhance the rates of dissolution of most oxide and sulfide ore minerals. Solution circulation within the system would result from the air-lift and convection effects of rising gas bubbles, convection currents caused by the variation in temperature between the top and bottom of the system, and pumping of solution in closed circuit within the system. Metal would be recovered from the pregnant solution in its circuit from the top to the bottom of the mine.

A similar technique could leach a particular area underground; only the leaching zone would need immersion. If the area constituted a worked-out part of the mine, bulkheads placed in appropriate drifts or openings would seal it off. If it were a fragmented or permeable area underground reached by boreholes, no seal would be required. Solution and/or gases would enter the leaching zone under pressure and either would be forced back to the surface by the internal pressure in the system or would require pumping. If the volume of the cavity leached were not too large, heating the influent leach solution could prove advantageous.

Another idea that may warrant consideration is that the flow of a low-amperage current could be directed from an electrode on one side of a

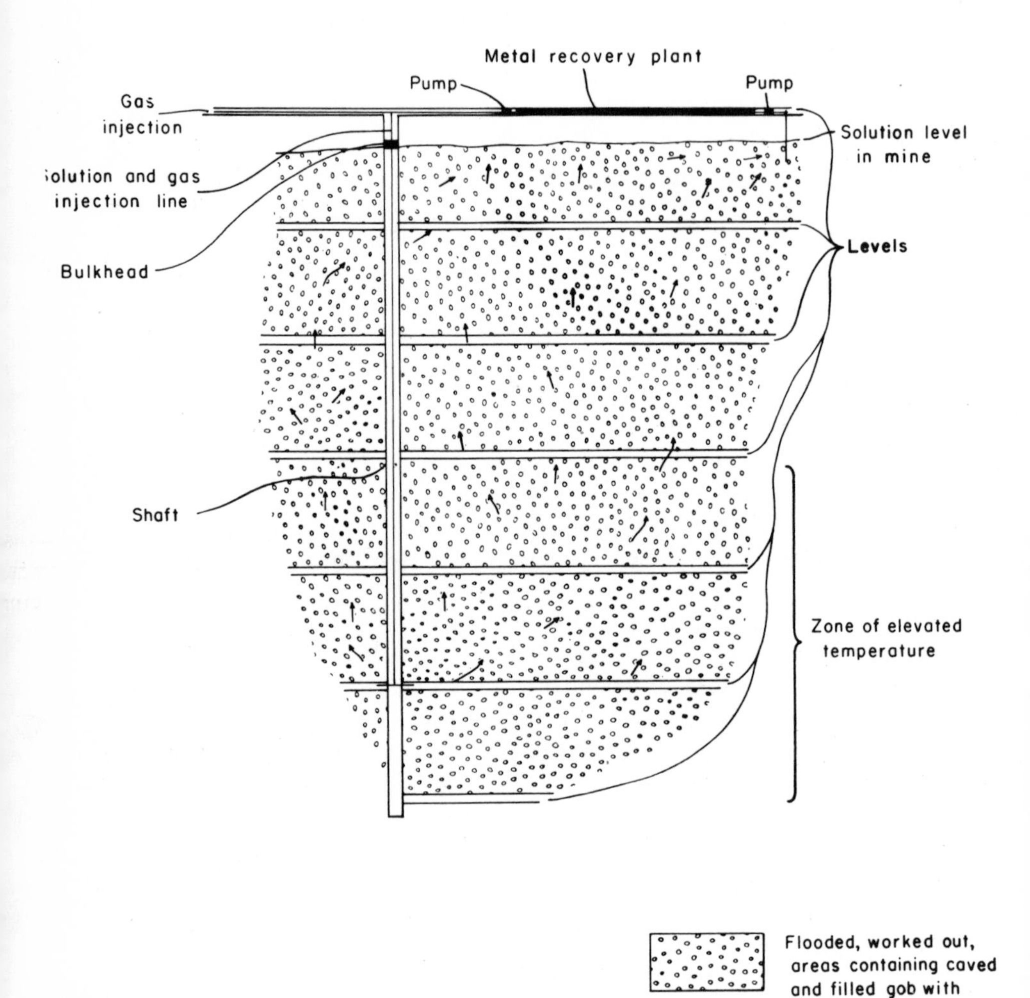

Figure 5.24. The leaching of an immersed mine under autoclave conditions (*courtesy of New Mexico Bureau of Mines and Mineral Resources*).

broken ore zone undergoing leaching to another electrode on the other side. Under given conditions of voltage and current, one might change the leaching solutions chemically, accelerate physicochemical reactions, and cause an increased rate of ion migration to a local area for solution recovery.

Inasmuch as the chemical reagents used in chemical mining generally constitute a major cost item and greatly influence leaching, reagent generation is a very important part of any chemical mining process. Some reagents can be generated and regenerated by natural processes in the leaching cycle, whereas others require various chemical processes.

Various industrial methods are used for producing and regenerating leaching solvents. Sulfuric acid is produced by the contact process; ferric sulfate can be made by the air oxidation of ferrous sulfates solutions in the presence of sulfuric acid, and each of the salts, acids, or bases used in leaching ($NaOH$, Na_2CO_3, $NaHCO_3$, $NaHClO_3$, and $NaCN$) is prepared commercially and can be regenerated from spent leach liquors. Unfortunately, the regeneration of these reagents is uneconomical in most instances.

Johnson[104] described a sulfuric acid-ferric sulfate solution generation and regeneration process for chemical mining. A specially designed, air-agitated autoclave generates sulfuric acid and ferric sulfate leach solutions from pyrite and concurrently oxidizes and hydrolyzes a spent ferrous sulfate leach liquor to sulfuric acid and ferric sulfate.

The key to low-cost lixiviants for chemical mining probably lies in the efficient use of either low-cost materials found in the ores to be leached, like pyrite, or raw or low-cost prepared materials readily available, like sodium chloride, trona, or liquid ammonia. Much research work needs to be done to determine how these materials can be used efficiently.

The last phase of any hydrometallurgical process, including chemical mining, is the recovery of metals from leach liquors. Conventional purification of a metal-containing solution followed by recovery of metals or compounds from the solution by either chemical or electrolytic precipitation is employed to obtain the marketable product. This recovery technique is adequately covered in the literature and its effectiveness is clearly demonstrated in several successful plant practices.

In connection with chemical mining applications, however, the recovery phase poses certain technical problems that may influence the over-all effectiveness of the process. One such difficulty concerns treating a large volume of very dilute metal bearing solutions that may contain more than one valuable metal. Unlike copper, not all metals easily precipitate on scrap iron. This may require recirculation of the leach solution to build up the metal content and then bleeding off of a small part of the concentrated leach stream for metal recovery.

Newer techniques of ion exchange, solvent extraction, and charcoal sorption may effectively concentrate dilute leach liquors. These procedures have proved very effective for processing large volumes of leach solutions containing more than one valuable metal. The Climax process[105] for recovering oxide molybdenum values by charcoal sorption and the New Mexico Bureau of Mines procedure[106] recently developed for recovery and selective separation of molybdenum, tungsten, and rhenium by sorption processes are typical of techniques that metal recovery systems will increasingly employ.

Since the crux of the chemical mining process is the particular lixiviant, any recovery phase that regenerates it or provides an essential component of the leach solution would be the preferable procedure. Also, since chemical mining depends on the continuous circulation of the leach solution at its peak volume, it is imperative that the retention time in the metal recovery step be as short as possible. This requirement necessitates a metal recovery procedure that is relatively quick and capable of handling large volumes effectively.

From the above discussion it is evident that chemical mining offers a novel approach for extracting metal values contained in low-grade deposits, worked-out mines, dumps, and tailing piles.

5.12 MINERAL RESOURCES AND THE ENVIRONMENT: NEED FOR BALANCE

In this chapter we have presented a brief but balanced introduction to the problem of man's relation to mineral resources. We have evaluated the consequence of the finiteness of mineral resources and examined the influence of varying social and economic factors on their utilization and adequacy. We have also seen that the limited resources available to man are being consumed rapidly, but that new sources are continuing to be made available through increased knowledge and improved technology. Finally, we have emphasized that the existence of any modern civilization depends on minerals and energy; however, the procurement of these needed resources is associated with certain pollution problems and disturbance of the environment.

It is obvious that we need to achieve a healthy balance between mineral resource extraction on one side and the environment on the other. Man's aim should be to attain economic growth such as to satisfy our needs and desires while at the same time preserving the environment. Through wise management of our mineral resources as well as the environment we can achieve a balance that is in the best interests of all.

To some degree, our environment can absorb the waste of civilization without harm and with a minimum of cost to society. However, pollution control becomes essential if the waste load imposed is so high as to endanger health, life, and environment. Control is also justified when the penalties imposed on the society from pollution are in excess of the cost of abatement. In both these cases, the major difficulty lies in measuring values. Therefore, it is imperative that environmental conservation practices and decisions involve the understanding and participation of all the parties concerned—government, industry, educators, scientists, and the public at large.

No industry in America today is more intimately involved in environmental controversy than the minerals-energy industry. It is true that the mining industry by its very nature contributes to certain environmental problems. However, the general public is usually unaware of the tremendous effort exerted by this industry to minimize the environmental impact of its operations. In addition, the industry has been faced lately with a very complex array of environmental regulations specified by the government agencies. The public fails to take into account the amount of time, energy and investment needed to achieve these regulations. They fail to realize that in many cases there are no instant solutions to environmental problems. For these reasons, it is essential that the minerals industry be given sufficient time to incorporate environmental programs which are technically sound and economically feasible. Moreover, the industry should be able to recover the added costs in their prices for the commodities that they produce.

The upswell of public opinion and government action concerning conservation of the environment compels mineral producers to seek new extractive processes that will achieve maximum conservation of finite supply of minerals with minimum contamination of the environment. This latter requirement is intimately related to the price structure of the commodity in question. It is apparent that in the past we have underpaid in dollars and overpaid in polluted environment for the goods and services we need. There is no doubt that in the future we will have to pay higher prices for metals and energy so that the damage to our environment may be kept to an acceptable level.

Technologically speaking, there are basically two solutions to the problem of environmental pollution created by mineral industries. The first short-range solution consists of adding environmental control to the existing plants as an appendage. Such an alternative will naturally increase the production cost which in turn will be passed on to the consumer. The second, long-range alternative constitutes the ultimate solution to the problem through research and development. Such efforts would result in totally new processes designed from the outset to eliminate or minimize the environmental problems. There are good possibilities that such innovative processes incorporating pollution controls, recycling provisions and by-products recovery will be comparable in cost to the old processes without such controls.

In order to achieve a healthy balance between mineral resources and environment in the future it is essential that the mining and utility companies do not develop a mineral resource unless it is economically feasible to incorporate proper restoration of the mined land and adequate pollution controls as integral parts of the original design as well as operation. Similarly, precautions should continue to be taken to protect the natural environment during the production and transportation of oil and gas through pipelines and ocean freight. Likewise, smelter and energy plant operators should provide positive pollution controls to prevent the escape of unwanted matter into the air through their smokestacks.

It is apparent that man has finally become keenly aware of both mineral shortages and environmental values, and he must now meet the challenge of extracting mineral wealth without creating environmental poverty. However, meeting this challenge is not going to be easy. On the one hand, it will require some very basic changes in philosophy as well as economics of the minerals industry. On the other, society will also need to change its philosophy concerning trade-off concepts and value judgement.

5.13 REFERENCES

1. Agricola, G. (1556) "De Re Metallica", Hoover Edition (1950), Dover Publications, Inc., New York, N.Y.
2. Brown, H., Bonner, J. and Weir, J. (1957) "The Next Hundred Years," The Viking Press, New York, N.Y.
3. Robie, E. H. (1964) "Economics of the Mineral Industries," American Institute of Mining Metallurgical and Petroleum Engineers, New York, N.Y.
4. McDivitt, J. R. (1965) "Minerals and Men," Resources for the Future, Johns Hopkins Press, Baltimore, Maryland.
5. Cloud, P. E. Jr. (ed.) (1969) "Resources and Man" by Committee on Resources and Man, National Academy of Sciences—National Research Council, W. H. Freeman and Co., San Francisco, California.
6. Ehrlich, P. R. and Ehrlich, A. H. (1970) "Population—Resources—Environment," W. H. Freeman and Co., San Francisco, California.

7. Flawn, P. T. (1970) "Environmental Geology," Harper & Row, New York, N.Y.
8. Detwyler, T. R. (1971) "Man's Impact on Environment," McGraw-Hill Book Co., New York, N.Y.
9. Lovering, T. S. (1969) "Resources and Man", 6—*Mineral Resources from the Land,* W. H. Freeman and Co., San Francisco, California, pp. 109–134.
10. Skinner, B. J. (1969) "Earth Resources," Prentice-Hall Inc., Englewood Cliffs, N.J. p. 10.
11. Mason, B. H. (1966) "Principles of Geochemistry," John Wiley and Sons, New York, N.Y.
12. Goldschmidt, V. M. (1954) "Geochemistry," Clarendon Press, Oxford.
13. Clarke, F. W. and Washington, H. S. (1924) "Composition of the Earth's Crust," U.S. Geological Survey Professional Paper 127.
14. Skinner, B. J. (1969) "Earth Resources," Prentice-Hall Inc., Englewood Cliffs, N.J. p. 17.
15. Cloud, P. E. Jr. (1968) *Realities of Mineral Distribution,* Texas Quarterly, volume II, pp. 103–126.
16. Barnett, H. J. and Morse, C. (1963) "Scarcity and Growth," Resources for the Future, John Hopkins Press, Baltimore, Maryland.
17. Weinberg, A. M. (1970) *Prudence and Technology—A Technologist's Response to Predictions of Catastrophe,* Philips Lecture, Haverford College, Philadelphia, October 7, 1970.
18. Wright, R. W. (1971) *Ferrous and Nonferrous Metal Resources,* "Centennial Volume," American Institute of Mining, Metallurgical and Petroleum Engineers, New York, N.Y. pp. 11–21.
19. Anonymous (1966) *Mining's Dynamic Age,* Engineering and Mining Journal, volume 167, June, pp. 131–142.
20. U.S. President's Materials Policy Commission, (1952) "Resources for Freedom," 5 volumes, William S. Paley, Chairman, Washington, D.C.
21. Foindbery, H. H. (1963) "Resources in America's Future," published for Resources of the Future, Inc. John Hopkins Press, Baltimore, Maryland.
22. U.S. Bureau of Mines, "Minerals Yearbook" published annually, Washington, D.C.
23. U.S. Bureau of Mines, "Minerals Facts and Problems," published every five years, 1960, 1965, & 1970, Washington, D.C.
24. Proctor, P. D. and Beveridge T. R. (1971) *Population, Energy, Selected Mineral Raw Materials, and Personnel Demand, 2000 A.D.,* Preprint Number 71-H-107, AIME Centennial Annual Meeting, New York, N.Y., March 1971.

25. Davis, W. B. (1971) *World Fossil Fuel Economics,* "Centennial Volume", American Institute of Mining, Metallurgical and Petroleum Engineers, New York, N.Y. pp. 42–61.
26. Skinner, B. J. (1969) "Earth Resources," Prentice-Hall, Inc., Englewood Cliffs, N.J. p. 122.
27. Cook, E. (1971) *Energy in Transition to the Steady State,* unpublished paper presented to American Association for the Advancement of Science, Philadelphia, Pa., December 30, 1971.
28. U.S. Bureau of Mines (1965) "Minerals Yearbook" and "Minerals Facts and Problems," Bulletin 650, 1970 Edition.
29. U.S. Bureau of Mines (1970) "Minerals Facts and Problems," Introduction: *The Evolving Minerals Economy,* Bulletin 650.
30. Tanabe, S., Takahashi, Y., and Iwasaki, I., (1967) *An Outlook of Raw Materials Supply for the Japanese Steel Industry,* 28th Annual Mining Symposium, University of Minnesota and Minnesota Section, AIME, January, 1967.
31. DeMille, J. B. (1947) "Strategic Minerals" McGraw-Hill Book Company, Inc., New York, N.Y.
32. Roush, G. A. (1939) "Strategic Mineral Supplies" McGraw-Hill Book Company, Inc., New York, N.Y.
33. Leith, C. K., (1943) "World Minerals and World Peace," Brookings Instritution, New York, N.Y.
34. Pehrson, E. W. (1945) *Our Mineral Resources and Security,* Foreign Affairs, volume 23, pp. 644–57.
35. Park, C. F. Jr. (1968) "Affluence in Jeopardy—Minerals and the Political Economy," Freeman, Cooper and Company, San Francisco, California.
36. Dobrin, M. B. (1960) "Introduction to Geophysical Prospecting" McGraw-Hill Book Company, Inc., New York, N.Y.
37. Harris, D. P. (1967) *Operations Research and Regional Mineral Exploration,* AIME, SME, Transactions, volume 238, pp. 352–58.
38. Peele, R. and Church, J. A. (1952) "Mining Engineers' Handbook," John Wiley and Sons, New York, N.Y.
39. Fletcher, J. B. (1971) *In Place Leaching—Miami Mine, Miami, Arizona,* Technical paper presented at Annual Meeting AIME, March, 1971, New York, N.Y.
40. MacGregor, R. A. (1969) *Uranium Dividends from Bacterial Leaching,* Mining Engineering, volume 21, March, pp. 54–55.
41. U.S. Department of the Interior (1967) "Surface Mining and Our Environment," Strip and Surface Mine Study Committee, Washington, D.C.

42. Cross, F. L. Jr. (1969) *Developments in Fluoride Emissions from Phosphate Processing Plants*, Journal Air Pollution Control Assoc. volume 19, January, pp. 15–17.
43. Donovan, P. P. (1969) *Lead Contamination in Mining Areas in Western Ireland*, Journal of Science, Food, and Agriculture, volume 20, January, pp. 43–45.
44. Lorenz, W. C. (1963) *Oxidation of Coal Mine Pyrites*, U.S. Bureau of Mines, Report of Investigation 6247, Washington, D.C.
45. Sheffer, H. W. and Evans, L. G. (1968) "Copper Leaching Practices in the Western United States," U.S. Bureau of Mines Information Circular 8341, Washington, D.C.
46. Grandt, A., (1969) "Operation Green Earth," Proceedings of Mining Environmental Conference, Univ. of Missouri, April 16–18, 1969, pp. 76–84.
47. Editor (1969) *Amax Anti-Pollution Efforts Win Recognition of Conservation Groups*, Engineering and Mining Journal, volume 170, May, pp. 116–118.
48. Newton, J. (1947) "An Introduction to Metallurgy," John Wiley and Sons, Inc. New York, N.Y.
49. Bray, J. L. (1947) "Non-Ferrous Production Metallurgy," John Wiley and Sons, New York, N.Y.
50. Kellogg, H. H. (1971) *Extractive Metallurgy in the Years Ahead—New Processes to Meet New Problems*, "A.I.M.E. Centennial Volume," New York, N.Y., pp. 147–160.
51. Strassburger, J. H. (1969) "Blast Furnace—Theory and Practice," Gordon and Breach Science Publishers, Inc., New York, N.Y.
52. Morgan, S. W. K. (1957) *The Production of Zinc in a Blast Furnace*, Trans. Institution of Mining and Metallurgy, London, England, volume 66, pp. 553–565.
53. Spedden, H. R., Malouf, E. E. and Prater, J. D., (1966) *Cone-Type Precipitators For Improved Copper Recovery*, Mining Engineering, volume 18, April, pp. 57–62.
54. Themelis, N. J. and Spira, P. (May 2, 1967) Canadian Patent 758,020.
55. Boldt, J. R., and Quaneuv, P. (1967) "The Winning of Nickel, Its Geology, Mining and Extractive Metallurgy." D. Van Nostrand Co., Inc. Princeton, N.J.
56. Forward, F. A. and Mackiw, V. N. (1955) *Chemistry of the Ammonia Pressure Process for Leaching Ni, Cu and Co from Sherritt Gordon Sulphide Concentrates*, Trans. A.I.M.E,. volume 203, pp. 437–463.
57. Dasher, J. and Power, K. (1971) *Copper Solvent-Extraction Process: from Pilot Study to Full-Scale Plant*, Engineering and Mining Journal, volume 172, April, pp. 111–115.

58. Dean, K. C., Dolezal, H., and Havens, R. (1969) *New Approaches to Solid Mineral Waste*, Mining Engineering, volume 21, March, pp. pp. 59–62.
59. Dean, K. C. and Havens, R., (1969) "Stabilization of Mineral Wastes from Processing Plants," Proceedings of Second Symposium, Mineral Waste Utilization Symposium, U.S. Bureau of Mines and IIT Research Institute, March 18–19, 1970, Chicago, Illinois.
60. Capp, J. P. and Spencer, J. D. (1970) "Fly Ash Utilization," U.S. Bureau of Mines, Information Circular 8483, Washington, D.C.
61. Humphrey, S. B. and Eikleberry, M. A. (1962) *Iron and Manganese Removal Using* $KMnO_4$, Water and Sewage Works 1962 Reference Number, pp. 68–70.
62. Ross, J. R. and George, D. R. (1971) "Recovery of Uranium from Natural Mine Waters by Countercurrent Ion Exchange," U.S. Bureau of Mines, Report of Investigation 7471, Washington, D.C.
63. Teller, A. J. (1971) *A Fresh Look at the Technology of Particulate Removal Via Scrubbing*, Engineering and Mining Journal, volume 172, April, pp. 81–82.
64. Ferguson, F. A. (1970) SO_2 *from Smelter; By-Product Markets a Powerful Lure*, Environmental Science and Technology, volume 4, pp. 562–568.
65. Argenbright, L. P. and Preble, B. (1970) SO_2 *from Smelters; Three Processes form an Overview of Recovery Costs*, Environmental Science and Technology, volume 4, pp. 554–561.
66. Swan, D. (1971) *Study of Costs for Complying with Standards for Control of Sulfur Oxide Emissions from Smelters*, Mining Congress Journal, volume 57, April, pp. 76–85.
67. Argenbright, L. P. (1971) *Smelter Pollution Control-Facts and Problems*, Mining Congress Journal, volume 57, May, pp. 24–28.
68. Devasto, P. J. (1971) *Effect of* SO_2 *Emission and Ambient Control on Copper Smelter Production and Operating Costs*, Preprint Volume, AIME Environmental Quality Conference, Paper No. EQC 62, New York, N.Y.
69. Bryk, P., Ryselin, J., Honkasalo, J. and Malmstrom, R. (1958) *Flesh Smelting Copper Concentrates*, Journal of Metals, volume 10, pp. 395–400.
70. Lorie, R. J. and Benson, B. (1968) *Pressure Hydrometallurgy Technology*, Paper presented at Annual AIME Meeting, New York, N.Y., February, 1968.
71. Haver, F. P. and Wong, M. M. "Recovering Elemental Sulfur from Nonferrous Minerals—Ferric Chloride Leaching of Chalcopyrite

Concentrate," U.S. Bureau of Mines, Report of Investigation 7474, Washington, D.C.

72. Staff (1967) *Copper Leaching with Cyanide—A Review of Five Inventions*, Engineering and Mining Journal, volume 168, September, pp. 123–124.
73. Bruynesteyn, A. (1970) *Microbiological Leaching—Research to Date and Future Applications*, Annual AIME Meeting, Denver, Colorado, February 18, 1970.
74. Norton, R. W. (1971) *California Oil Industry's Solution for Oil Spill Control in the Santa Barbara Channel*, Preprint Volume, AIME Environmental Quality Conference, Paper No. EQC 10, New York, N.Y. pp. 45–50.
75. Nelson, T. C. (1925) *Effects of Oil Pollution on Marine Life*, Appendix V to the Report of the U.S. Commission of Fisheries for 1925, Bureau of Fisheries Document, pp. 171–181.
76. *National Environmental Policy Act* (NEPA), Act of January 1, 1970, Public Law 91-190 (83 Stat. 852).
77. *Water Quality Improvement Act of 1970*, Title 1 of Public Law 91-224, Act of April 3, 1971 (84 Stat. 91).
78. *Refuse Act, 33 Code of Federal Regulations* (CFR) Part 209, Permits for Discharge of Deposits into Navigable Waters, Federal Register, volume 76, no. 67, April 7, 1971, pp. 6564–6570.
79. *Clean Air Act Amendments of 1970*, 42 U.S.C. 1957, as Amended by Publications 91-604, December 31, 1970.
80. *Spills of Oil and Hazardous Substances*, Title 1 of Public Law 91-224, approved April 3, 1970 (84 Stat. 91), 33 CFR Part 153 and 18 CFR Part 610.
81. Dole, H. M. (1970) *Keynote Address—Solid Wastes: A Natural Resource*, Proceedings of the Second Mineral Waste Utilization Symposium, IIT Research Institute, Chicago, Illinois, pp. 3–4.
82. Reno, H. T. and Brantley, F. E. (1970) "Iron," Minerals Facts and Problems, U.S. Bureau of Mines Bulletin 650, pp. 291–314.
83. Spendlove, M. J. (1970) *A Profile of the Nonferrous Secondary Metals Industry*, Proceedings of the Second Mineral Waste Utilization Symposium, IIT Research Institute, Chicago, Illinois, pp. 88–92.
84. Anonymous (1966) "Standard Classification for Nonferrous Scrap Metals," National Association of Secondary Materials Industries, Inc., Circular NF 66.
85. Sullivan, P. M. and Stanczyk, M. H. (1971) "Economics of Recycling Metals and Minerals From Urban Refuse," U.S. Bureau of Mines Technical Progress Report-33, Washington, D.C.

86. Chindgren, C. J., Dean, K. C., and Peterion, L. (1971) "Recovery of Nonferrous Metals from Auto Shredder Rejects by Air Classification," U.S. Bureau of Mines Technical Progress Report-31, Washington, D.C.
87. Kuznetsov, S. I., Ivanov, M. V., and Lyalikovo, N. N. (1963) "Introduction to Geological Microbiology," McGraw-Hill Book Co., Inc., New York, New York.
88. Colmer, A. R. and Hinkle, M. E. (1947) *The Role of Microorganisms in Acid Mine Drainage,* Science, volume 106, pp. 253–256.
89. Colmer, A. R., Temple, K. L., and Hinkle, M. E. (1950) *An Iron-Oxidizing Bacterium from the Acid Drainage of Some Bituminous Coal Mines,* Journal of Bacteriology, volume 59, pp. 317–328.
90. Bryner, L. C., Beck, J. V., Delmar, B. D., and Wilson, D. G. (1954) *Microorganisms in Leaching Sulfide Minerals,* Industrial and Engineering Chemistry, volume 46, pp. 2587–2592.
91. Trussell, P. C. and Duncan, D. W. (1966) *Recent Advances in the Microbiological Leaching of Sulfides,* Preprint No. 66B21, Annual AIME Meeting, New York, New York.
92. Malouf, E. E. and Prater, J. D. (1961) *Role of Bacteria in the Alteration of Sulfide Minerals,* Journal of Metals, volume 13, pp. 353–356.
93. Sutton, J. A. and Corrick, J. C. (1961) "Possible Uses of Bacteria in Metallurgical Operations" Information Circular 8003, U.S. Bureau of Mines, Washington, D.C.
94. O'Connor, L. T., Brierley, J. A. and Bhappu, R. B. (1971) *Bacteria Leaching Process,* New Mex. Inst. of Min. and Tech., Socorro, New Mexico, Patent Pending.
95. Fisher, J. R. (1966) *Bacterial Leaching of Elliot Lake Uranium Ore,* Transactions, Canadian Institute of Mining, volume 69, pp. 167–171.
96. Johnson, P. H. and Bhappu, R. B. (1969) *Chemical Mining—Theoretical and Practical Aspects,* State Bur. of Mines and Mineral Res., New Mex. Inst. of Min. and Tech., Socorro, New Mexico.
97. Lasky, S. G. (1950) *How Tonnage and Grade Relations Help Predict Ore Reserves,* Engineering and Mining Journal, volume 151, April pp. 81–85.
98. Levine, N. M. and Hassialis, M. D. (1962) *The Physical Factors Governing the Leaching of Ores,* Preprint No. 62B80, Annual AIME Meeting, New York, N.Y.
99. Johnson, G. W. (1959) *Mineral Resource Development by the Use of Nuclear Explosives,* UCRL-5458, University of California, Lawrence Radiation Laboratory, Livermore, California.

100. Griswold, G. B. (1968) *Rock Fracturing Techniques for In Place Leaching*, Annual AIME Meeting, New York, N.Y.
101. Robie, E. H. (1967) *New Method Suggested for Leaching Uranium Ore In Place*, Engineering and Mining Journal, volume 168, May, pp. 106–107.
102. Pirson, S. J. (1959) *Better Phosphate Mining is Suggested*, Engineering and Mining Journal, volume 160, June, pp. 208–210.
103. Davis, J. G. and Shock, D. A. (1968) *Solution Mining of Thin Bedded Potash*, Annual AIME Meeting, New York, N.Y.
104. Johnson, P. H. (1965) *Acid—Ferric Sulphate Solutions for Chemical Mining*, Mining Engineering, volume 17, August, pp. 65–68.
105. Staff, (1966) *Climax Dedicates Acid Leach-Charcoal Absorption Process Moly Oxide Ores*, Mining Engineering, volume 18, December, pp. 88–91.
106. Reynolds, D. H., Long, W. V., and Bhappu, R. B. (1966) *Select Separation of Molybdenum, Tungsten, and Rhenium by Sorption Processes*, Proceedings, Second Annual Operating Conference of Metallurgical Society, AIME, Philadelphia, Pa., December.

Chapter 6

SOLID WASTE DISPOSAL AND REUSE

6.1 INTRODUCTION

Environmental control is a field of science and engineering created by this age of advanced technology. It encompasses the control and treatment of man-made wastes, such as garbage, scrap, and tailings. These seemingly undesirable materials may be the means of creating new industries and of augmenting our supply of raw materials. In general, pollution and waste are considered as problems; however, it is high time that we look at these discards as important sources of metals and raw materials.

The idea that an individual or corporation can dispose of waste without creating a greater nuisance than can be tolerated by an urban society was proven unsound long ago. As villages grew into cities, the disposal of waste inevitably became a function of local government. In addition, since most people had no interest in refuse once it was beyond their sight and smell, the government usually took the cheapest choice, namely, to concentrate the nuisance by dumping the unwanted material at a selected location. The first criterion for selecting a method of waste disposal was that it be inexpensive because few people wanted to spend money on materials they considered worthless. Increased knowledge of environmental factors related to public health and improved sanitary standards have helped to extend the criteria for acceptability beyond cost considerations. In addition, urban sprawl has reduced the number of acceptable dumping sites in or near where the waste is discarded to almost the vanishing point.

Several estimates have been made of the amount of waste produced daily in the United States.[1–4] This is a difficult figure to estimate and varies according to which sources are included in the estimate. One estimate is 10 pounds per day per capita.[3,4] A breakdown of the 10 pounds per day per capita estimate shows that an average of 5.32 pounds (Table 6.1)[4] of solid wastes were collected by municipalities nationwide. The remaining 4.68 pounds were disposed of by private concerns or individuals. The 10 pounds estimate includes household, industrial, and commercial wastes; but does not include approximately 7 million passenger cars, trucks, and buses which are junked annually.

TABLE 6.1.

Average solid waste collected in the United States[4] (pounds per person per day)

Solid wastes	*Urban*	*Rural*	*National*
Household	1.26	0.72	1.14
Commercial	0.46	0.11	0.38
Combined	2.63	2.60	2.63
Industrial	0.65	0.37	0.59
Demolition, construction	0.23	0.02	0.18
Street and alley	0.11	0.03	0.09
Miscellaneous	0.38	0.08	0.31
Totals	5.72	3.93	5.32

Waste includes liquids, solids, and gases. It refers to useless, unwanted, unused, or discarded materials. Industrial fumes and smoke constitute most of the gases. The liquids of municipal waste consist mainly of sewage and the fluid part of industrial wastes. The solids of municipal waste are classed as refuse. Refuse includes garbage, rubbish, ashes, street refuse, dead animals, abandoned vehicles, industrial wastes, demolition wastes, construction wastes, special wastes, and sewage treatment residue (Table 6.2).[5]

Garbage decomposes rapidly and may produce objectionable odors. When it is not properly stored it is food for vermin and a breeding place for flies. Garbage has some value as animal food, as a base for commercial animal feeds, for its grease content, and as a plant fertilizer after processing.

Organic combustible rubbish is not highly putrescible and may be stored for relatively long periods without becoming a nuisance. Its high heat value when dry, permits incineration without forced draft and without added fuel. Noncombustible rubbish is not burnable at ordinary incinerator temperatures (1300–2000°F). Under ordinary conditions food left in tin cans desiccates rather than putrefies, and therefore cans are usually classified as rubbish rather than garbage.

Refuse has been analyzed and subdivided in another manner for various sections of the United States (Table 6.3).[6] The percentage content of the components is somewhat different for different areas. The percentage of combustible material has increased as more consumer goods have been packaged and intended for shorter life spans.

Municipal governments generally do not consider themselves as obligated to collect industrial refuse, but rather consider industry obligated for its disposal. Because industrial refuse may endanger public health, its storage, hauling and disposition are subject to municipal control. Many local governments are now providing for disposal of this type of refuse. When an industry contracts to a private company for disposal of wastes it is still generally held accountable. Laws vary from state to state, but it is generally impossible to contract away liability. Therefore, when industry produces dangerous wastes which require special handling, industry must take care of their own disposal.

The different methods of waste disposal commonly employed are briefly

TABLE 6.2.

Refuse materials classified by kind, composition, and sources[5] (courtesy of Public Administration Service)

Kind	*Composition*	*Sources*
Garbage	Wastes from preparation, cooking, and serving of food; market wastes; wastes from handling, storage, and sale of produce.	Households, restaurants, institutions, stores, markets.
Rubbish	*Combustible:* paper, cartons, boxes, barrels, wood, excelsior, tree branches, yard trimmings, wood furniture, bedding, dunnage. *Noncombustible:* metals, tin cans, metal furniture, dirt, glass, crockery, minerals.	
Ashes	Residue from fires used for cooking and heating and from on-site incineration.	
Street refuse	Sweepings, dirt, leaves, catch basin dirt, contents of litter receptacles.	Streets, sidewalks, alleys, vacant lots.
Dead animals	Cats, dogs, horses, cows.	
Abandoned vehicles	Unwanted cars and trucks left on public property.	
Industrial wastes	Food processing wastes, boiler house cinders, lumber scraps, metal scraps, shavings.	Factories, power plants.
Demolition wastes	Lumber, pipes, brick, masonry, and other construction materials from razed buildings and other structures.	Demolition sites to be used for new buildings, renewal projects, expressways.
Construction wastes	Scrap lumber, pipe, other construction materials.	New construction, remodeling.
Special wastes	Hazardous solids and liquids: explosives, pathological wastes, radioactive materials.	Households, hotels, hospitals, institutions, stores, industry.
Sewage treatment residue	Solids from coarse screening and from grit chambers; septic tank sludge.	Sewage treatment plants; septic tanks.

TABLE 6.3.

Expected ranges in mixed refuse composition[6]

Component	*Percent composition as received (dry weight basis)* Anticipated range	Nominal
Paper (Total)	37–60	55
Newsprint	7–15	12
Cardboard	4–18	11
Other	26–37	32
Metallics (Total)	7–10	9
Ferrous	6–8	7.5
Nonferrous	1–2	1.5
Food	12–18	14
Yard	4–10	5
Wood	1–4	4
Glass	6–12	9
Plastic	1–3	1
Miscellaneous	<5	3
TOTAL		100
Moisture content:		
Range (Percent) 20–40		
Nominal (Percent) 30		

described in this chapter. The conditions under which they are employed and the problems encountered are also mentioned. Recycling of wastes is considered from both an economic and a technical viewpoint.

6.2 RECYCLING WASTES

Currently, considerable effort is being made to utilize a wide variety of waste materials, ranging from abandoned automobiles to scrap metals, paper mill sludges, slags and tailings, and municipal refuse. The primary motivations in this recycle trend are profit and environmental considerations. Industry has always conducted waste utilization programs in the form of by-product recovery wherever possible, because of profit incentive. However, today the profit motive is not the only criterion for practicing waste utilization. A second look at waste utilization is warranted any time when an accumulation of waste becomes an environmental problem or when the cost of disposal is unreasonably high.

An equally compelling reason for considering waste utilization is the growing realization that critically needed resources for the present as well as the future are wasted. As we have seen earlier for the case of mineral resources (See Chapter 5), the consumption of metals, nonmetals, and fossil fuels is increasing at an astonishing rate. We will need to resort to each and every available source of minerals to satisfy our needs and wants. For these reasons it is imperative that we develop new technology to make marketable by-products from waste in order to reduce disposal costs as well as to broaden our resource base.

According to Dole,[7] in 1968, about 300,000 tons of aluminum were utilized in the manufacture of cans, lids, and caps which under the existing disposal practices will never be reclaimed. Nearly 25,000 tons of tin (used as a coating on cans) are discarded in our city dumps annually while our country is importing this valuable commodity. One may be forced to wonder why we consider these items as "solid waste" when the resource potential of such discards is so obvious.

Currently, about 50 percent of the total supply of copper in the United States is recovered from scrap, while for many years more lead has been recovered from scrap than from domestic ores. Today, the secondary metals business is a vital activity amounting to about eight billion dollars and playing an important role in the economic life of the nation. Moreover, secondary materials move in international business both as exports and imports, and play a vital role in the economic growth and stability of developing nations. Without such resources, these countries would suffer serious industrial and economic handicaps. Many larger countries in Europe and in the Far East are regular buyers of scrap generated in this country, and as a result of this international trade, the United States gains in balance of payments.

Secondary metals are those recovered from waste materials such as scrap metals while *primary* metals are derived directly from natural ores. Scrap may be derived from *capital goods* such as railroads, bridges, buildings, and industrial machinery, or from *consumer goods*, such as automobiles, refrigerators, and cooking utensils. The supply of scrap has a very marked effect on the price of primary metals. When business is good, price levels for scrap are more or less determined by the supply and demand for primary metals. However, when consumption is at a low level, scrap exerts a depressing influence on the price. In general, the price of secondary metals is lower than that of primary ones, due to the presence of impurities in scrap. This is not the case, however, if the scrap is processed in primary smelting along with ores and concentrates.

Out of more than an estimated 100 million tons of new iron and steel consumed yearly in the United States,[8] about 25 percent reverts to scrap, and is subsequently reprocessed on a 20-year cycle. From 1964 through

TABLE 6.4.

United States secondary metal consumption*

Metal	*Quantity consumed 1967*	*Units*	*1969 Value[1] million dollars*
Copper	1,243,000	Tons	1,096
Aluminum	885,000	Tons	403
Lead	554,000	Tons	161
Zinc	263,000	Tons	76
Antimony	25,568	Tons	26
Tin	22,790	Tons	71
Mercury	22,150	Flasks	12
Silver	59,000,000	Ounces	101
Gold	2,000,000	Ounces	70

* Bureau of Mines Office of Mineral Resource Evaluation.
[1] Prices, American Metal Market, May 7, 1969.

1968, consumption of scrap by the iron and steel industry averaged 87.8 million tons per year, compared with an average of 88.9 million tons of pig iron during the reference period. In dollar value, gross sales of iron and steel scrap alone amounted to about $3 billion in 1970. Naturally, automobile scrap is the single largest source of scrap with a contribution of about 6–7 million tons annually. Railroad scrap is the second largest with 3.5 million tons.

The role of secondary metals in the nonferrous industry is equally important both from economics and conservation standpoints. Spendlove[9] has suggested that if it were not for the existence of the nonferrous secondary-metals industry, copper would be accumulating as junk at a rate exceeding 1.25 million tons, aluminum at 885,000 tons, lead at 554,000 tons, and zinc at 263,000 tons. Table 6.4 indicates the importance of nonferrous secondary metals in the United States.

The most economical materials to purchase for reclamation are the ones which require minimum processing. Most of the lead recovered as secondary metal is obtained from discarded automobile batteries, pipes, type metal, solder, and shot.[10] The reclamation of lead and antimony from old batteries has become a very important business in the last 30 years with more than 50 percent of the total secondary lead obtained from this single source. Secondary zinc is largely recovered from brasses, die castings, zinc sheet trimmings and the like. Most of the reclaimed copper is derived from copper wire, utensils, bearings, faucets, pipe, and automobile radiators. Secondary aluminum is obtained mostly from utensils, castings, clippings, and machine shop scrap.

Since scrap materials constitute a large variety of pure as well as alloyed heterogeneous mixtures, classification and sorting of scrap metals constitute important phases of secondary metal processing. Several types of portable devices have been designed to identify metals. The *metalscope*, using the principle of a spectrograph; and the *sortometer*, operating on the thermo-electric reading generated at the heated junction of two dissimilar metals, are but two examples of recent developments in the field.

Two of the major problems encountered in processing scrap are to completely utilize all values from the plant effluent (such as fumes, dust, and slags), and to safely remove unwanted impurities from alloy melts. The important consideration here is not to create additional pollution problems in trying to solve an existing one.

Efforts by the United States Bureau of Mines[11] to separate, recover, and recycle materials contained in a large variety of mineral-laden solid wastes are aimed at eliminating the current waste of valuable resources found in urban refuse and municipal incinerator residues. Such residues contain approximately 75 percent by weight of metals and glass. A workable method using screening, shredding, grinding, magnetic separation, and gravity concentration techniques was developed by the United States Bureau of Mines to treat municipal incinerator residues and produce metallic iron concentrates, clean nonferrous metal scrap, clean fine glass fractions, and fine carbonaceous ash as tailings. The estimated capital costs for such treatment plants vary from $1.4 million for a 250 tons per day capacity plant to $1.7 million for a 1,000 tons per day plant; while the operating cost was estimated for the same size plants at $4.06 and $1.09 per ton of dry residues, respectively. On the other hand, the estimated value of recovered constituents amounted to about $15 per ton of residue treated. From these figures it is obvious that urban refuse and municipal incinerator residues can be treated at a profit *if markets can be developed.*

Another important contribution from the Bureau's laboratories is the processing of automobile scrap to recover valuable metals.[12] Besides containing iron and steel, such refuse includes millions of dollars worth of nonferrous metals which are discarded annually. Each ton of automobile scrap contains about $50 worth of copper, zinc, aluminum, and lead and it is estimated that about $15 million worth of these metals is discarded annually. The recovery system includes shredding of automobile bodies, followed by magnetic separation to remove ferrous metals. The remaining nonferrous values are next recovered by an air classification system to separate the lighter non-metallic fragments from the metallic fraction. Work is continuing to develop processing techniques capable of separating individual nonferrous metals from each other, for final recycling.

6.3 METHODS OF DISPOSAL

There are several methods of disposal of solid wastes. Approximately 90 percent of such wastes are dumped on land, 8 percent are incinerated, and less than 2 percent are used for feeding hogs.

Dumping without preparation of offensive waste, either to people or to the environment, on the ground or in the ocean requires no description. This type of dumping, which has been very common in the past, should not be continued. It is wasteful of land and water. Open dumps provide places for diseases, vectors, and vermin to multiply, and become injurious to man and other animals. As has been noted earlier, indiscriminate dumping on bodies of water can destroy aquatic life as well as make the beaches unsightly, unhealthy, unpleasant, and unuseable. Leachate from waste can contaminate ground water or surface water, and also make them unsightly, unhealthy, unpleasant, and unuseable. Natural processes may clean the water if not overpowered and allowed sufficient time and space.

Sanitary Landfill

Sanitary landfill is the most widely used and generally least expensive method of refuse disposal in the United States. Refuse deposited in a sanitary landfill is compacted and covered to prevent odors, fires, rodents, insects, and other nuisances usually associated with open dumping. Careful planning and control may result in the efficient disposal of all components of refuse. A sanitary landfill need not pollute water or air. The land occupied by the landfill can, and should, be reclaimed (Figure 6.1).

It is easier to use filled land for recreational parks without structures because such land is unstable for many years after filling. Sanitary landfills have been used for agriculture, residences, parking lots, and airfields. Several-storied structures have been built on them when careful attention has been paid to special engineering problems created by drainage, uneven settling, and the generation of foul-smelling explosive gas. Uneven settling and bearing capacity can be somewhat improved by separating the types of refuse during construction of the landfill.

Selecting an area for a sanitary landfill is a process in which many factors must be considered. The site should be located as close to the col-

Figure 6.1. This sanitary landfill is being built in a coal strip mine. This system accomplishes two purposes in that it removes the refuse and the strip mine from sight. The reclaimed land may later be used for an industrial park.[13]

Figure 6.2. These refuse collection trucks are lined up at a centrally located transfer station.[13]

Figure 6.3. This is the lower side of the refuse transfer station shown in Figure 6.2. The refuse is compacted into these trailers by hydraulic rams before the trailers are hauled to the sanitary landfill shown in Figure 6.1.[13]

Figure 6.4. The compacted refuse is unloaded from the trailers by equipment in the trailer.[13]

Figure 6.5. A bulldozer spreads and compacts refuse after discharge from transfer trailers.[13]

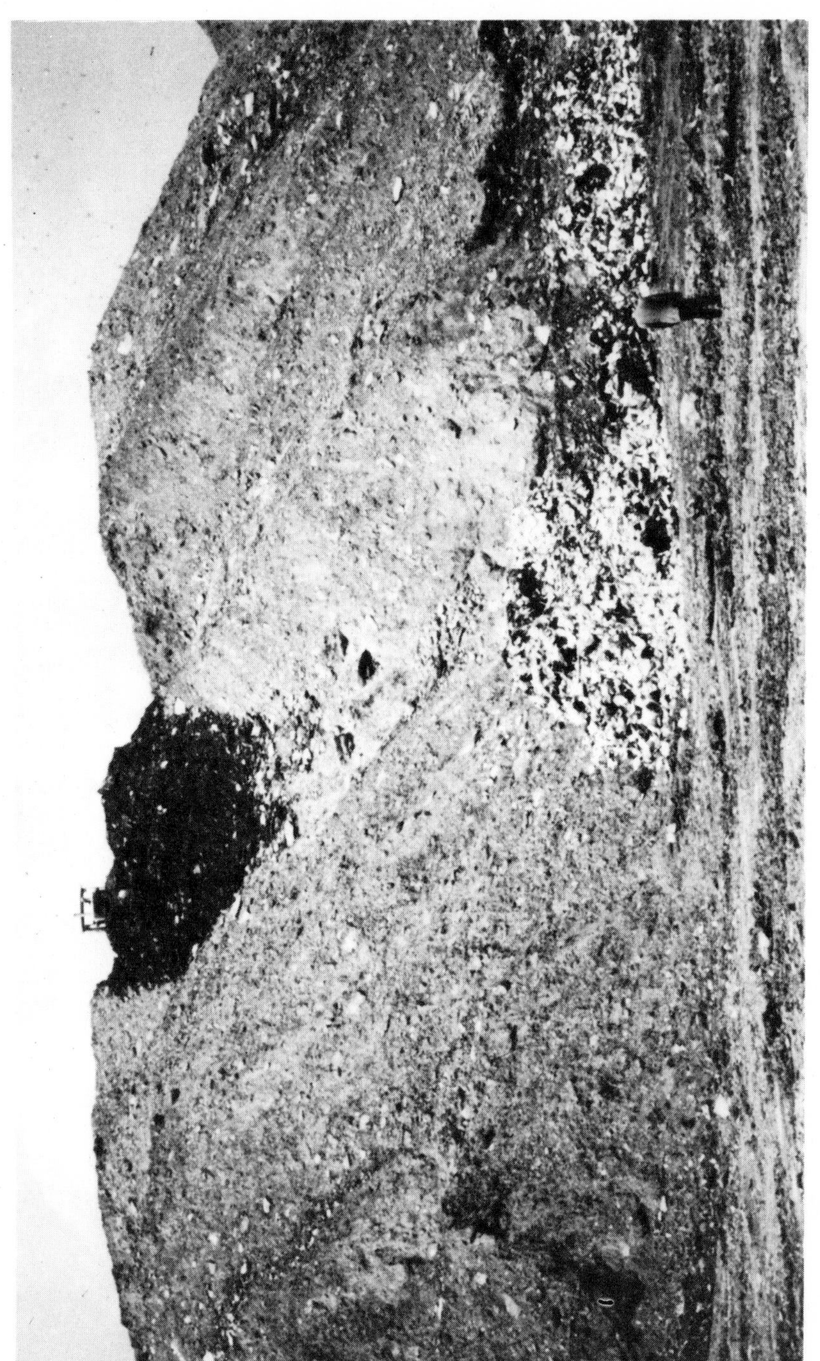

Figure 6.6. The bulldozer covers compacted refuse with soil bank material left over from strip mining.[13]

lection area as possible because haulage costs from the point of household pickup to a landfill constitute 80 percent of overall disposal costs. When the collection area is far from the disposal site, transferring the refuse to larger highway vehicles becomes the more economical method of operation (Figures 6.2–6.6). Approximately 1 acre of land is required for 10,000 persons for 1 year of operation if the depth of the refuse is compacted in 7-foot lifts.[13] Consideration should be given to location so that prevailing winds will not blow toward the community served.

Earth cover material should be evaluated for quality and quantity. Sandy loam free of rocks larger than 6 inches is recommended as cover material.[5] Experience indicates that earth containing less than 35 percent coarse fragments is satisfactory. The material after compaction must preclude ponding of water. It must prevent vectors and small animals from leaving or entering the refuse. A minimum of 6 inches of material is required for daily cover. Each 7-foot lift should be covered by at least another foot of earth. The final cover for the landfill should be over 2 feet thick (Figure 6.6).

The prevention of groundwater pollution should be considered in the selection of a site. Pollution of groundwater will occur when the site is over or adjacent to an aquifer; or when the water in the fill from precipitation, from decomposition, or from an artificial source saturates the refuse to above field capacity so that leached fluids are produced, and the leachate is capable of entering an aquifer. In a humid climate it will be almost impossible to prevent the above conditions, and therefore, the site must be prepared so that the leachate will not enter any aquifer. A foot of undisturbed earth should be left between the landfill and the water table for each foot of fill depth.

Gas and water samples collected from drilled test holes in and near landfills can be used to determine the extent of pollutional leachate formation. Local topography and drainage patterns can be used to estimate the quantity of water expected to flow through the refuse. The geology of the area will determine if the leachate will enter the rocks forming the boundaries of the landfill.

It is important to know the geology of the area to be used in a landfill operation. Leachate should filter through soil before reaching bedrock, a condition that is precluded if the refuse contacts bedrock. The chemical activity, structure, and fracture patterns of the rocks need to be examined. If the structure is too complex it may be impossible to predetermine the direction of leachate flow. When the chemical activity of the rocks is too high, the leachate may enlarge fractures by solution and thus develop channels to aquifers. After the channels are developed there will be little renovation of the leachate en-route to the aquifers.

The following steps are recommended in preparing a landfill site[5,13]·

1. Make a study to permit accurate planning for access roads, drainage, lift heights, diversion channels, dikes, levees, soil characteristics, and life of the site.
2. Build a semipermanent, all-weather road on the site and a vehicle turn-around if needed.
3. Build an all-weather access road to the site.
4. Take measures to prevent paper from being windblown.
5. Build appurtenances, such as an earth berm or solid fence, around the site to screen the activity for aesthetic reasons.
6. Provide suitable facilities for storing and servicing equipment.
7. Provide facilities for workmen.
8. Install scales to weigh refuse to help determine costs and to improve management practices.
9. Contour the surface of the fill to provide for immediate water runoff in order to eliminate ponding or washout.
10. Place 6 feet of cover material (sandy loam with less than 35 percent coarse fragments) between the refuse and bare rock to prevent their coming into contact. Clay should be used to cover any coal seams.
11. In cases where leachate may enter ground water, install a drainage and collection system before the landfill is constructed. The bottom of the landfill should be covered with an impermeable barrier and then covered with permeable earth material, such as gravel, which will drain via pipes to a collection and treatment system. Since the refuse may produce leachate for over 20 years, the treatment system should be designed for long term operation.

The ramp method of operating a landfill in a strip mine is commonly used (Figures 6.1, 6.4–6.6). The ramp-type operation is especially desirable because the heavy moving equipment is continuously driven over the face of the fill working area. A ramp having a maximum slope of 30° from the horizontal is made from high ground to low ground. The refuse may be dumped at the top or the bottom of the ramp. (Better compaction is achieved when the refuse is dumped at the bottom of the lift and spread upwards to form a cell). Refuse is spread and compacted in about 2-foot layers, after which it is covered with at least 6 inches of earth. These cells are built up to form a lift about 7 feet high. Each lift is covered with at least a foot of earth before another lift is built on top of it. The final lift should be covered with at least 2 feet of earth and immediately seeded to prevent erosion.

Another method of constructing a landfill is containerizing the refuse and then unloading the container at the landfill (Figure 6.7). The con-

Figure 6.7. A special trailer hauls two refuse containers from transfer station to landfill site where a modified power shovel removes containers from the trailer.[13]

Figure 6.8. A refuse container is moved into position above the hopper of a compacting machine known as "the mole".[13]

Figure 6.9. Refuse is discharged into the hopper of compactor; the cover of the container prevents paper from being windblown.[13]

Figure 6.10. A piston compresses refuse in the bottom of the hopper, while a telescoping cover follows the piston.[13]

Figure 6.11. A bulldozer pushes loose fill dirt over compacted refuse that has been extruded from the rear of the in-place compactor.[13]

tainers are unloaded into a machine which compacts the refuse in a predug trench (Figures 6.8–6.11). Then a bulldozer covers the compacted refuse in a manner similar to the ramp method.

Adequate earthmoving equipment should be provided for the excavating of drainage systems, moving and compacting the refuse, placing and compacting the earth cover material over the refuse, and finishing to maintain the desired grade and contour of the fill. The type and size of the dozers and scrapers depends on the method of operation, and characteristics of the site. A 35,000-pound crawler tractor with a dozer blade can generally dispose of 300 tons of refuse in an 8-hour working day by using the ramp method.

Inspection for and control of rodents, should be maintained on a continuous basis. All ponded surface water should be drained and the depression filled with non-putrefying material. During fill operations, an insect spray should be applied when necessary. All voids on fill bands and surface should be filled and compacted.

Monitoring the activity of the landfill may be accomplished in several ways. Gas generated within the refuse may be bled-off with vertical pipes spaced throughout the fill. Often sufficient gas is evolved to flare. As the rate of decomposition decreases, the quantity of gas should decrease. An increase in moisture in the refuse tends to increase the rate of decomposition. The settling rate of the fill may be measured by surveying bench marks around the fill and installing observation markers in the fill during construction. Generally most of the settling occurs during the first 2 years after construction. Settling and decomposition will vary according to the content of the fill.

Ground-water pollution from the fill can be monitored by drilling test holes down to the water table and obtaining water samples. Lysimeters should be installed beneath landfills to determine if the landfill seal is adequate and permanent. Surface water drainage should also be sampled. The samples should be analyzed for chemical and biological pollution. Chemical (chlorine) contamination from a landfill has been detected in wells over 1,000 feet away from it.

Because of the population explosion and its concentration in urban areas, the demand for land and unpolluted water is increasing along with the quantity of solid waste. These factors point to the urgent need for methods of disposal that require less land area and result in less or even no pollution. Cost is an important factor in the efficient design and management of a disposal technique. Moving waste material away from population centers to allow for natural renovation is sometimes politically and economically unfeasible. Studies on landfill hydrology indicate that a way of solving the leachate problem is to design landfills so that leachate can be collected and renovated before it reenters the hydrologic system. There is an urgent need to design an economical system to handle the landfill leachate. Because leachate is produced for many years after the landfill is in place, the treating system should have some other purpose as well; such as treating sewage from a nearby population center.

Incineration

Incineration is the most common method of reducing the volume of refuse and changing it into a less toxic or dangerous form. This is essentially a method of rapid oxidation, but the high heat maintained in municipal incinerators also melts many noncombustibles, and alters the chemical form of others.

Chemical alteration is not always favorable such that toxic substances may be formed or previously non-water soluble substances may become soluble pollutants. Therefore, some caution must be utilized in selecting the materials to be incinerated. However, most refuse can be incinerated safely such that the method is widely used.

Incineration can provide an economic, nuisance free, sanitary method of refuse disposal. However, it may or may not be the most satisfactory solution to the disposal problems of a particular industry or city. Its advantages and disadvantages must be evaluated according to the local conditions prevailing.

For a city the advantages of central incineration include:

1. Reduction in volume such that much less land is required than for sanitary landfills handling unprepared refuse. Transportation costs are reduced because of the great decrease in weight and volume.
2. The incinerator ash residue can contain negligible amounts of organic material and thus be almost nuisance free. The ash is usually acceptable as fill material.
3. Site selection is not as difficult as in some other disposal methods such that a central location for an incineration plant is possible. When an attractive building is built with well-landscaped grounds the plant is acceptable in many neighborhoods.
4. Incinerators are little affected by climate.
5. A modern incinerator can consume most combustible refuse and also reduce the volume of much of the noncombustible material. However, it is limited to relatively small objects. Materials which produce excessive smoke or are explosive are not acceptable.
6. The capacity of an incinerator is adjustable by changing the hours of operation and extent of combustion.
7. Some income may be obtained from incinerators in the form of

heat and salvageable metals. Industry may be willing to buy the services of the municipal incinerator rather than maintain its own disposal method.

Some of the disadvantages of incineration to be compared with its advantages are as follows:

1. Municipal incinerators require a large capital investment (Figure 6.12). Costs of construction range from about $4,000 per ton capacity to over $8,000 per ton capacity depending upon type of construction. The higher capital investment is related to waste heat recovery by steam generation. When the steam can be sold for a sufficiently high price the higher initial cost is justified by the income produced.
2. Operating costs for incineration are higher than landfill operation because of the wages of the more skilled employees required to operate, maintain, and repair the incinerator. Moreover maintenance and repair costs are higher than for landfills because of damage to the furnace caused by the high temperatures maintained and type of material burned. Refuse that contains wire, tramp metal, abrasives, explosives, and fusible material frequently damages equipment and machinery. The higher operating costs may be offset by lower transportation and land costs compared to the sanitary landfill method.
3. Regardless of an attractive building and landscaping, refuse disposal operations are offensive to many people such that site location may be difficult. Heavy truck traffic in the vicinity of the plant may be considered a hazard and nuisance in some neighborhoods.
4. Recently enacted pollution control regulations may require prohibitively expensive equipment for compliance. Figure 6.12 does not take these expenses into account.
5. Residue and fly ash from the burning process must be disposed of by other means than incineration. Incinerators cannot dispose of all types of waste material such as demolition wastes and special wastes. Explosive materials such as naptha and sawdust create a hazard for personnel and may damage the furnace such that it must cease operation for repairs. Other means of disposal for these wastes must be maintained. When the incinerator is not operational other means of disposal for all wastes must be readily available.

Determining the incineration plant location and size is both an engineering and a social problem. The social problem may be the more difficult. Selecting a centrally located site will reduce hauling costs to the

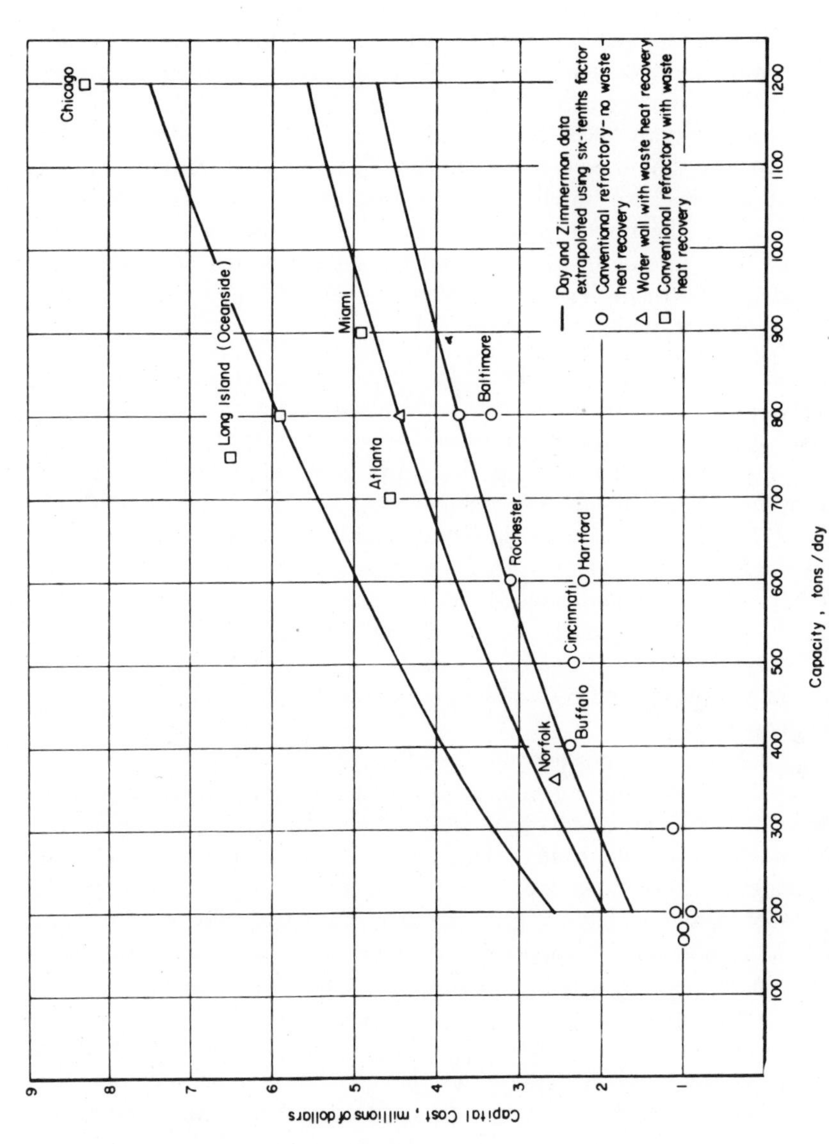

Figure 6.12. Capital costs of municipal incinerators.[6,14]

plant, but if material is to be salvaged the nearest shipping point may also be important. When determining the size of the plant one should consider:

1. *The quantity and type of refuse to be burned.*
2. *The hours of operation planned.* One, two, or three 8-hour shifts are common. Continuous operation is more efficient and reduces damage of the refractory material by reducing thermal shock.
3. *Seasonal peak loads and additional capacity for maintenance.* The capacity should be between the highest and lowest refuse collection periods because building for the peak refuse day will mean that much of the time the capacity is not used. Storage capacity may be built to contain the overload for a short period of time. Plans should be made for expansion as the quantity of refuse increases along with the population growth.

Just as with sanitary landfills scales should be provided to weigh the refuse for purposes of planning, cost accounting, and as a means of billing if the plant is shared with other municipalities or industry.

The tipping floor where the trucks dump the refuse into the storage pit must be large enough for easy maneuverability and should extend the length of the pit. The floor should have a highly impervious surface which is easy to clean and resistent to abrasion and shock.

The storage pit should have sufficient capacity to take care of emergencies. Depending upon the hours of operation and stand-by capacity, the pit may have from 12 to 36 hours of the incinerator's burning capacity. Refuse in the pit will putrefy quickly such that some means of odor suppression are necessary.

The combustion of refuse progresses through stages of drying, igniting, burning of solid wastes, burning gaseous wastes, and deposition of particulate matter before discharging the gaseous end products into the atmosphere. Most of the burning occurs in the furnace, which is a refractory lined structure containing grates and supplied with large volumes of air. A more complete burning of the airborne particles occurs in what may be called the combustion chamber. This is a large insulated chamber where exhaust gases are expanded to reduce their velocity. The reduced gas velocity permits particulate matter to settle out.

The completeness of the burning is largely affected by the retention time of the combustible material in the incinerator's high-temperature zones and the quantity of air available. Baffles are built in the combustion chamber to improve the mixing of the hot material with the air. Additional air is added above the flames to increase the available oxygen for the particulate matter and burnable hot gases.

The residue from a well operated incinerator ranges from 5 percent by

Figure 6.13. Test cell is covered with canvas that is tied down to wooden planks around the top edge of the cell.[13]

Figure 6.14. Flexible hose from airblower to test cell delivers enough air to support canvas and control burning.[13]

weight for refuse relatively free of noncombustibles to 25 percent for mixed refuse. This is equivalent to 3 to 15 percent by volume. In properly burned residue the unburned organic matter will range from 2 to 7 percent by weight.

High temperature burning destroys odors and reduces noxious gases to acceptable levels. Of course what is termed acceptable varies according to the situation. The furnace should maintain a temperature between 1300° and 2000°F. and average 1700° to 1800°F. The minimum is required to decompose aldehydes, mercaptans, and other malodorous hydrocarbons. Above the maximum temperature refractories start to deteriorate rapidly.

Because of the shortage of land suitable for landfill construction near areas of solid waste generation and the value of land near areas of high population density, efforts are being made to reduce the volume of refuse sent to landfills and to accelerate landfill stabilization. Several methods are available, some of which have already been mentioned. Reducing the quantity of material discarded by changing living habits is perhaps the most difficult method of reducing volume. Incineration and its variations reduce the volume and sometimes produce a recycleable product. Compacting refuse at the landfill site is the most commonly practiced method of volume reduction. Compacting and baling the refuse has several advantages such as ease of handling, transportation, and higher densities.

One of the most innovative processes has been developed by Ralph Stone and Company.[13] This process amounts to low temperature incineration at the landfill site without the high capital investment required to build a conventional incinerator (Figures 6.13–6.14). A pit is prepared with walls made of sandy loam, through which the exhaust gases from the burning refuse are filtered. The rate and temperature of burning are controlled by the input of air. A cell with a capacity of 118 cubic yards of refuse was "incinerated" in about 20 days. Volume reduction was nearly 90 percent. Costs were estimated to be $323,000 to build and operate a facility to handle 100,000 tons of refuse per year.

Summary

Solid waste disposal will continue to be a problem for many years in the future regardless of the great deal of effort being put into developing means of recycling wastes. For instance, the majority of the mixed refuse consisting of food and papers cannot be reused in its present form. Some of the organic material can be treated to make petroleum type products. Composting is partially successful, as is using the heat produced from incineration. Under present marketing conditions about 10 percent of the waste metals and glass are considered reusable. Of course these quantities may increase as the technology of using contaminated (mixed) raw materials improves.

The monetary value of waste material is greatly increased if it is not mixed. Perhaps a reasonable solution to the mixing occurring during waste collection will be found, but until an economic incentive is created to recycle waste, disposal will remain a problem.

Differences in local economies produce markets for different products, but the market from any single product manufactured from waste material is unlikely to support a waste recovery system. Therefore, a system producing several products is desirable. However, increased product streams increases capital investment and decreases product quantity in each stream. This is particularly disadvantageous for small processing plants.

Several waste recovery systems have been suggested[14] which combine processes. Each is directed toward a specific objective, but includes other salvage operations. All the systems include salvaging steel, tin, iron for precipitation of copper, material suitable for manufacture of structural objects, and residual material for landfills. Most systems include as specific goals waste heat recovery (steam), biochemical conversion (alcohol and protein), and pyrolysis (carbon). Selecting the appropriate objective will depend upon local conditions.

6.4 REFERENCES

1. Witt, P. A. (1971) *Disposal of Solid Wastes,* Chemical Engineering, volume 78, pp. 62–69.
2. Hoog, H. (1970) *Conserving Our Environment, What Chemical Engineering Can and Should Do,* Presidential Address, Annual General Meeting of the Institution of Chemical Engineers, April 21, 1970.
3. American Chemical Society (1969) "Cleaning Our Environment—The Chemical Basis for Action," American Chemical Society, Washington, D.C.
4. Black, R. J., Muhich, A. J., Klee, A. J., Hickman, H. L., and Vaughan, R. D. (1968) "The National Solid Waste Survey; An Interim Report," U.S. Department of Health, Education and Welfare, Cincinnati, Ohio, 53 pp. (Paper also presented at a meeting of American Public Works Association.)
5. American Public Works Association (1966) "Municipal Refuse Disposal" 2nd ed., Public Administration Service, Chicago, Illinois.
6. Drobny, N. L., Hull, H. E., and Testin, R. F. (1971) "Recovery and Utilization of Municipal Solid Waste," U.S. Environmental Protection Agency, Publication No. SW10C.

7. Dole, H. M. (1970) *Keynote Address—Solid Wastes: A Natural Resource*, Proceedings of the Second Mineral Waste Utilization Symposium, IIT Research Institute, Chicago, Illinois.
8. Reno, H. T. and Brantley, F. E. (1970) *Iron*, "Mineral Facts and Problems," U.S. Bureau of Mines Bulletin 650, pp. 291–314.
9. Spendlove, M. J. (1970 *A Profile of the Nonferrous Secondary Metals Industry*, Proceedings of the Second Mineral Waste Utilization Symposium, IIT Research Institute, Chicago, Illinois, pp. 88–92.
10. National Association of Secondary Materials Industries, Inc. (1966) "Standard Classification for Nonferrous Scrap Metals," Circular NF 66.
11. Sullivan, P. M. and Stanczyk, M. H. (1971) "Economics of Recycling Metals and Minerals From Urban Refuse," U.S. Bureau of Mines Technical Progress Report 33.
12. Chindgren, C. J., Dean, K. C. and Peterion, L. (1971) "Recovery of Nonferrous Metals from Auto Shredder Rejects by Air Classification" U.S. Bureau of Mines Technical Progress Report 31.
13. Sheffer, H. W., Baker, E. C., and Evans, G. C. (1971) "Case Studies of Municipal Waste Disposal Systems," U.S. Department of the Interior, U.S. Bureau of Mines Information Circular 8498.
14. Day & Zimmerman, Associates (1968) "Special Studies for Incinerators," Public Health Service Publication No. 1748, Washington, D.C.

Chapter 7

THE ATMOSPHERE

7.1 INTRODUCTION

The earth is enveloped by a gaseous mass in which all terrestrial forms of life develop. This envelope is the outermost of the geochemical spheres and consists of a mixture of gases and water vapor.

For many centuries man has considered the atmosphere as an unrestricted receptacle for the product of his many activities. It was not until the past few decades that man began to understand and worry about the ecological effects of pollution of the environment. As shown in previous chapters, many efforts have been made and continue to be made in order to abate and control pollution. The government is working closely with universities and research institutes to solve many specific pollution problems. In addition, industry is contributing much time, money, and effort to reduce stack emissions, clean-up the air, reduce automotive pollution and in a general sense make the air more breathable and the land more livable.

Because of the great importance of preserving the quality of the air, this chapter will be devoted to giving a general description of the most important characteristics of air pollution and related aspects. We will attempt to describe in some detail the main air pollutants and their sources, some of the photochemical processes that take place in the atmosphere, and the effects of air pollution on man and the environment. A brief discussion of the meteorological parameters that control atmospheric processes will also be given. Finally, consideration will be given to what we can expect in the future, as far as air pollution is concerned. For a more detailed analysis of the atmosphere and atmospheric pollution the reader is referred to more specialized references.[1–8]

The basis for the entire problem of air pollution is that people do not take the atmosphere as a valuable resource but rather consider it simply as something that is there. It is very difficult, as pointed out in Chapter 1, to put a value to the individual and collective usage of the atmosphere. This failure on man's part to manage this resource has led directly to the problem of air pollution that we presently face.

The problem of air pollution is intensified in our large cities due to the congestion of factories, automobiles, and people in relatively small areas of land. Air pollution in our urban areas is the result of the inability of atmospheric dilution processes to take care of the large amounts of chemicals dumped into the atmosphere.

7.2. PHYSICOCHEMICAL NATURE OF THE ATMOSPHERE

As a whole, the atmosphere can be described in terms of its physical constitution and its chemical composition.

Air, for all practical purposes, may be considered as a perfect gas composed of molecules and atoms set in random motion. The atmosphere contains numerous charged particles (electrons and ions) under the influence of the earth's electric and magnetic fields. Most of the atmosphere is concentrated close to the earth's surface (half of it is confined within the first 5.5 km (3.4 mi)), yet it extends up to 1000 km (620 mi), and beyond.

There are some constituents that do not greatly vary in the atmosphere both in time and space. Oxygen and nitrogen account for up to 99% of the air (by volume), and the relative concentrations of these gases do not change with altitude up to 90 km (56 mi). Besides these constituents, one of the most important is water vapor which may be present in amounts varying from 0.02 to 4.0% by weight. This amount depends on several factors, of which temperature is the most significant. Water vapor plays an important role in regulating climatic conditions.

There are also minor constituents of the air that affect the physicochemical nature of the atmosphere. Some of these constituents will be treated in this chapter.

Composition

The air we breathe is composed of a rather simple mixture of gases. These can be grouped into:

(a) The group of permanent constituents of which two account for 99% of the total atmospheric volume;
(b) The group of inert constituents; and
(c) The group of variable constituents.

Table 7.1 lists both the permanent and inert constituents of air while Table 7.2 lists some of the variable constituents. Besides the gaseous constituents listed in these tables, an enormous number of suspended solid and liquid particles of different sizes ranging from 5×10^{-3} to 20 microns in radius are present in the atmosphere. These particles are called *aerosols*

TABLE 7.1.

Principal constituents and inert gases that make up air at ground level.

Molecule	*Molecular mass*	*Percentages by volume*
N_2	28.014	78.084
O_2	31.999	20.946
A	39.948	0.934
CO_2	44.009	0.033
Ne	20.183	0.0018
He	4.0026	0.00053
Kr	83.800	0.0001

and play a most important role in controlling the mechanism of droplet condensation and in controlling the formation of ice crystals. Many aerosols are involved in the chemical processes that take place in polluted atmospheres.

One of the most important properties of air is the constancy of the relative proportions of the permanent gases up to a height of about 90 km

TABLE 7.2.

Some of the minor constituents present in the atmosphere.

Molecule	*Atmospheric region where present**
O_3 (ozone)	its concentration increases with height and reaches a maximum in the stratosphere
H_2O (water vapor)	the water vapor concentration is quite variable in the first 10 km and dissociation occurs in the mesosphere
NO_2, NO (oxides of nitrogen)	concentration of these oxides is quite variable in the troposphere, and is greatly affected by industrial stack effluents and photochemical processes; NO_2 dissociates in both the mesosphere and stratosphere
CO (carbon monoxide)	concentration of this constituent is variable, especially in the troposphere
CH_4 (methane)	variable concentration in the troposphere; oxidation occurs in the stratosphere and dissociation in the mesosphere
SO_2 (sulfur dioxide)	released into the atmosphere by volcanoes, plant decay, and as a result of fuel consumption

* For a definition of each of the atmospheric regions mentioned above see the text and Figure 7.1.

(56 mi). Convection currents and mixing processes maintain the relative proportions of these gases up to this height. Above it, the processes of diffusion and diffusive equilibrium begin to take place. Also above 90 km (56 mi) the proportion of lighter gases increases with height, and diffusion becomes more important than mixing. In addition, photodissociation with production of atomic oxygen begins to occur and must eventually affect the composition of the upper atmosphere above 90 km (56 mi).

Atmospheric Regions

The atmosphere has been divided for study purposes in different ways. One of the most common ways is based on differences in temperature. Figure 7.1 shows the regions in which the atmosphere has been divided according to temperature and the specific phenomena that take place at different heights.

The lowest region of the atmosphere is called the *troposphere* and is characterized by a decrease in temperature with increasing altitude. The troposphere ends in the *tropopause.* This is a layer of the atmosphere where no change in temperature with altitude is observed. The tropopause occurs at varying altitudes depending on the latitude and season.

Above the tropopause extends the *stratosphere* up to 50 km (31 mi), and in this region temperature increases with altitude. The *mesosphere* is a region situated above the *stratopause* layer in which the stratosphere ends. The mesosphere extends from 50 ± 5 km (31 ± 3 mi) up to the *mesopause* at 85 ± 5 km (53 ± 3 mi) in altitude. In the mesosphere temperature decreases rapidly with height and very important photochemical processes occur. As a result of such processes, airglow emissions are observed.

The region above the mesopause is the *thermosphere,* extending from 85 km (53 mi) up into outer space. In this region temperature increases with altitude. Temperature gradients here are the result of absorption of energy, that is, ultraviolet radiation with a wave length less than 1750 Å is gradually absorbed by the atmospheric gases and a large part of the absorbed energy is used to heat the thermosphere.

Atmospheric Composition Nomenclature

Considering its composition, the atmosphere can be divided into two zones:

(a) The homosphere. This region extends from ground level up to the mesopause (Figure 7.1). Here the relative proportion of the principal constituents of air does not change. This implies the same mean molecular mass.

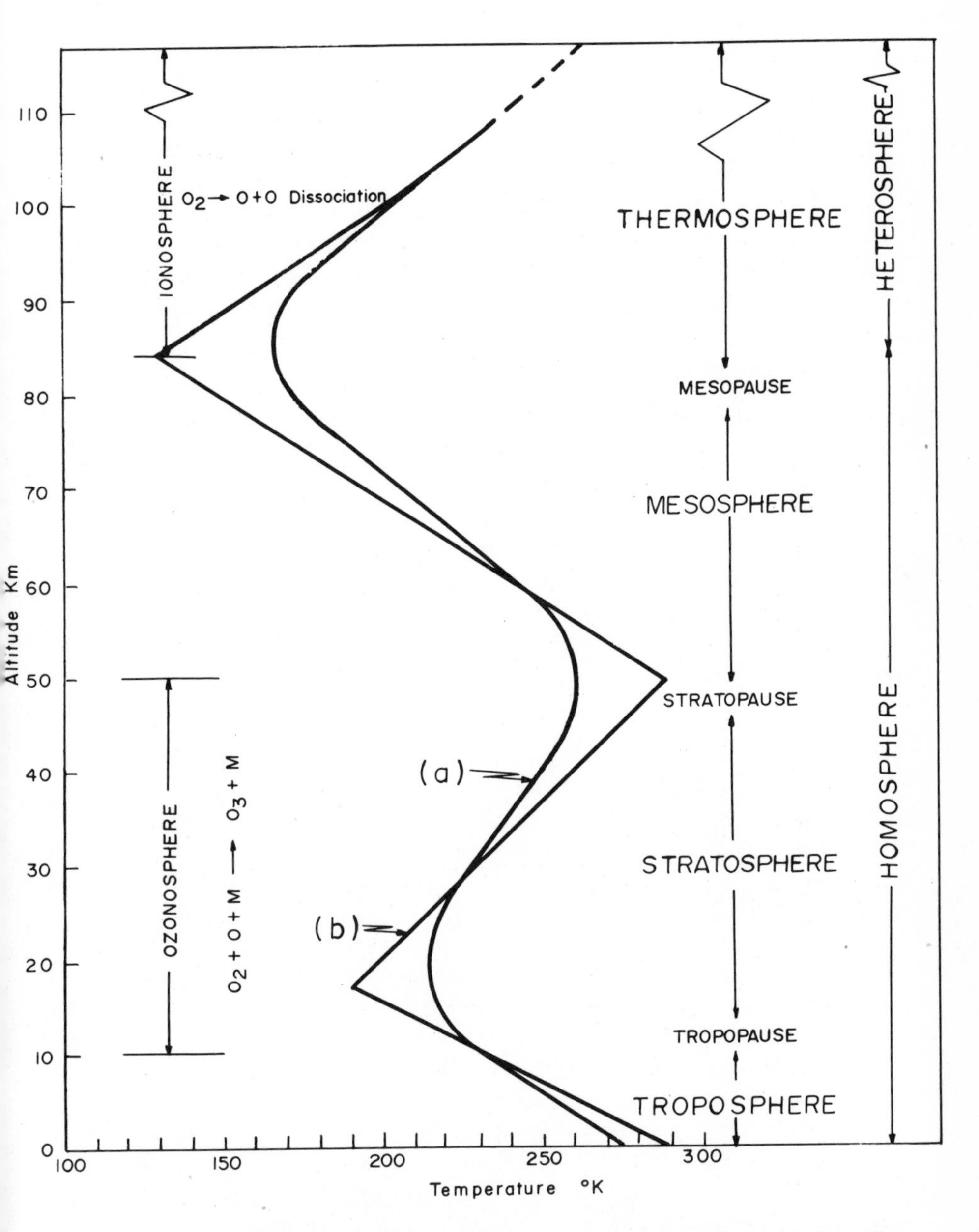

Figure 7.1. Atmospheric regions and atmospheric nomenclature. Curve (a) shows the profile of temperature with altitude as expected for average conditions, while curve (b) corresponds to a theoretically calculated model.

(b) The heterosphere. This region extends from the mesopause into outer space. Here air composition begins to change with height due to the dissociation of oxygen and diffusion processes.

If some specific atmospheric constituents are considered, the region of their occurrence may be named according to these constituents. For example, the *ozonosphere* is the region where ozone formation and destruction take place. It is located at a height of 10–50 km (6–31 mi). The *ionosphere* has been defined[9] as the part of our planet's outer atmosphere where ions and electrons exist in quantities large enough to affect the propagation of radio waves. It usually extends from 50 km (31 mi) to great heights.

Molecular Behavior of Gases

Atmospheric gases obey the equation of state for an ideal gas (at least in an approximate fashion). From kinetic theory a simple equation of state for an ideal gas can be deduced assuming that a gas consists of molecules which have their mass concentrated in a very small spherical volume. The molecules are assumed to collide with each other and with the walls surrounding them without loss of kinetic energy. It is also assumed that molecules exert no forces on each other except when they collide. With all the above assumptions included, the equation of state can be written as follows:

$$PV = nRT \tag{7.2-1}$$

where

P = pressure,
V = volume of gas,
T = absolute temperature in °K,
n = number of moles, and
R = Specific Gas Constant = 8.314×10^7 ergs·mole^{-1}·°K^{-1}

The above expression gives only an adequate equation of state useful to solve a series of practical problems involving the group of permanent atmospheric gaseous constituents. Water vapor and other substances found in air need a more elaborate equation of state.

The equation of state for an ideal gas involves a system in equilibrium and predicts that at constant pressure a volume of an ideal gas varies linearly with temperature. The constant of proportionality R is the same for all ideal gases. The thermodynamic properties of such a gas can be applied with good accuracy to unsaturated dry air; however, when air is saturated with water vapor, important variations from ideal gas behavior occur. For such a case, the idealized assumptions that molecules occupy a negligible volume individually or as a whole, and that there are no attraction forces between them, cannot hold and a modification has to be made to the equation of state. Since molecules occupy a finite volume and there are attraction forces between them, these attractive forces increase or decrease depending on the molecular separation. There are strong repulsive forces when the separation between molecules is about one molecular diameter.

In 1873 van der Waals deduced an equation of state taking into account both corrections mentioned in the preceding paragraph. He stated that when pressure increases or temperature decreases, the volume must approach a small but finite value "nb" and therefore V in equation (7.2-1) should be replaced by $V - nb$. In addition, he represented the pressure at a point within the gas as the sum of the pressure exerted by the walls and the pressure exerted by the intermolecular attraction.

The van der Waals' equation of state is written as:

$$\left(P + \frac{an^2}{V^2}\right)(V - nb) = nRT \qquad (7.2\text{-}2)$$

where "a" and "b" are constants which depend on the gas' properties.

Solar Radiation and Atmospheric Absorption

The sun is the source of virtually all the energy for all the processes that take place in the earth-atmosphere system. The sun's radiant energy is emitted as electromagnetic rays traveling through free space at a speed of 300,000 km/sec (186,000 mi/sec). Such rays cover a wide range of wave lengths and their totality is called the *electromagnetic spectrum*. For the purposes of our discussion, three regions of the spectrum are of utmost importance, namely, the ultraviolet, visible, and infrared. These three regions account for 99% of the total solar radiation.

Incoming solar radiation is attenuated by three main processes, namely, reflection, molecular scattering, and absorption. These three processes occur mainly in the troposphere where the densest region of the atmosphere is located.

Oxygen and nitrogen account for 99% of the air (by volume) and are responsible for the strong absorption of wave lengths shorter than 3000 Å.* This absorption plays a very important role in controlling the photochemical processes that take place in the upper atmosphere.

The absorption of the effective ionizing radiation is confined to the wave length range below 1300 Å, and dissociation of oxygen occurs for wave

* 1 Å = 1 angstrom = 10^{-4} microns = 10^{-10} meters.

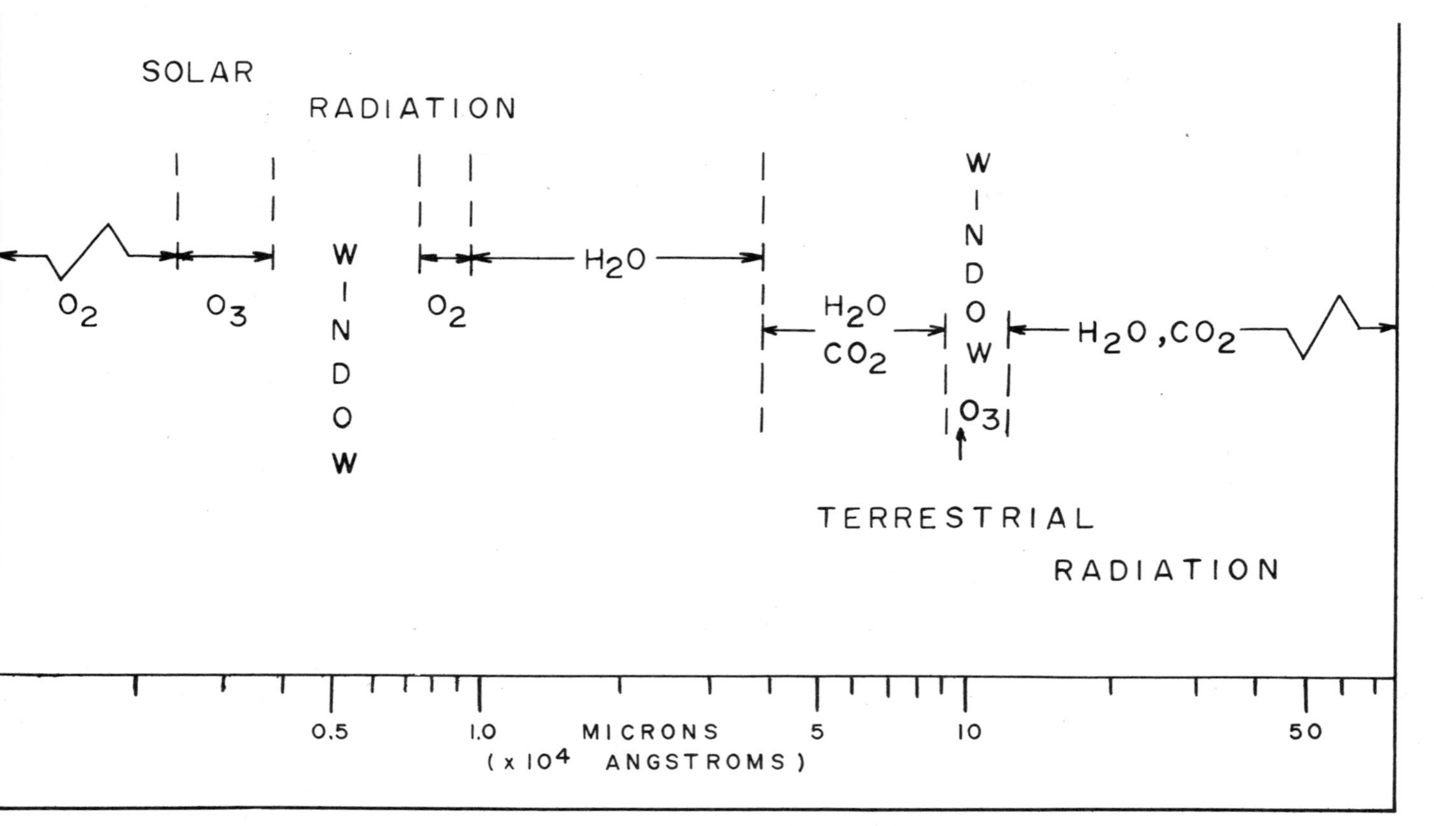

Figure 7.2. Absorption of solar radiation by earth's atmosphere. (From ref. No. 10).

lengths smaller than 2000 Å. Wave lengths of lesser energy excite molecules and atoms to resonance-transition states giving sharp atomic absorption lines or molecular bands. Ultraviolet radiation is absorbed by ozone in the ozonosphere.

Most infrared radiation is absorbed by water vapor and carbon dioxide. The atmosphere is nearly transparent for wave lengths ranging from 3000 Å to 8000 Å where the visible range of the spectrum is located. Within the 8–12 micron range the atmospheric window is located, which is the interval of the strongest outgoing terrestrial radiation, except at 9.6 microns where a strong ozone-absorption wave length is found. Figure 7.2 shows the wave lengths where absorption is significant in the atmosphere.[10]

7.3 THE AIR WE BREATHE

The average man when traveling can think that the atmosphere is different in composition at different places. One observes a clear blue sky over the oceans and a thin black strip of polluted air in some crowded cities.

He soon realizes that the air over oceans, deserts, mountains, and farmlands is clear compared with the air in urban areas, and becomes aware that such a big difference in the air is mostly a consequence of man-made pollution. It is presumed today that air pollution began when our ancestors became organized in fixed communities and started using fire for cooking, heating, and tool making. As population and human activities increased in concentrated geographical areas, the gaseous and particulate pollution increased in some direct relationship to fuel consumption, population growth, and refuse burning.

Centuries later, coal replaced wood as a source of energy. The adaptation of such fuel for industrial uses and domestic heating culminated in Europe during the industrial revolution. It became evident since that time that the consumption of large quantities of the "unnatural" fuel as it was called produced the first observed symptoms of *dangerous* substances put into the air. This fact caused the formation of "action committees" to protest against the use of coal.

Today, man lives in such an industrialized age that all life-supporting environments have been severely altered. Every year man pours millions of tons of toxic substances into the air. Carbon monoxide from auto-exhaust and fossil-fuel consumption adds up more than 58% of the total atmospheric gaseous pollutants. Sulfur oxides from fuel combustion, sulfuric acid plants, and smelters comprise the second most abundant group of pollutants in the atmosphere of the United States. Hydrocarbons and oxides of nitrogen from automotive emissions and fuel combustion also

TABLE 7.3.

Estimated nationwide emissions (1968).[11]

	10^6 tons/year				
Source	*Carbon monoxide*	*Partic-ulates*	*Sulfur oxides*	*Hydro-carbons*	*Nitrogen oxides*
Transportation	63.8	1.2	0.8	16.6	8.1
Fuel combustion in stationary sources	1.9	8.9	24.4	0.7	10.0
Industrial processes	9.7	7.5	7.3	4.6	0.2
Solid waste disposal	7.8	1.1	0.1	1.6	0.6
Miscellaneous	16.9	9.6	0.6	8.5	1.7
Total....................	100.1	28.3	33.2	32.0	20.6

play an important role in the photochemistry of polluted air and are virtually responsible for the occurrence of *photochemical smog* (see section 7.6). Besides, there is a broad variety of other contaminants from a great number of emitting sources some of which are not man-made (such as volcanic emissions, some forest fires, etc). Table 7.3 lists five major pollutants emitted by man's activities and the work of nature in the United States during 1968.[11]

It should be noted, however, that not all the pollutants that are dumped into the atmosphere stay as such in a stagnant state. Most are diluted by atmospheric diffusion processes. Some react with others and still others settle back onto the land. If it were not for the occurrence of natural processes that remove pollutants from the atmosphere the air of our large cities would have already become unbreathable.

In summary, it can be said that the "ordinary" air that man breathes in most big cities is contaminated with gaseous and particulate pollutants from fossil fuel consumption, transportation, factories, burning of domestic and municipal refuse, forest fires, and a wide variety of sources. Of course, such pollution is the result for the most part of man's desire for the fruits of industrialization. A big question can be raised at this point, namely, can we have industralization without excessive pollution? The answer to this question must be yes if civilization is to survive and prosper.

7.4 PARTICULATE MATTER

Particulate matter can be defined as any material (except uncombined water) that exists as a solid or liquid in the atmosphere or in a gas stream, under given standard conditions. These conditions are included in the definition of particulate matter because some compounds are gaseous at stack conditions but are condensed in the ambient atmosphere.[12]

All particulate matter suspended in the air is subject to numerous physical and chemical phenomena. One of the most obvious properties of particulate matter is their settling ability. Particles settle in the air following the physical principle known as Stokes' Law. This law says that the settling velocity of a spherical particle in air is given by the equation:

$$v = \frac{gd^2(\rho_1 - \rho_2)}{18\eta} \text{ cm/sec} \qquad (7.4\text{-}1)$$

where

g = acceleration of gravity (cm/sec^2),
d = particle diameter (cm),
ρ_1, ρ_2 = density of particle and air respectively (gm/cc), and
η = viscosity of air in poises.

The above law cannot describe the settling rate of all particulate matter. For non-spherical particles corrections are needed. Also the law does not apply to particles smaller than 50 microns. This is because the atmosphere is a dynamic system and eddy and convective currents can extensively modify the extremely small gravitational motion described by equation (7.4-1).

Particulates have different modes of motion depending on their size. As has been mentioned, particles bigger than 50 microns have a settling motion according to Stokes' Law. Particles in the range from 1 to 50 microns approximate predictions of this law for the most part. On the other hand, particles smaller than 0.1 microns in diameter are mostly subject to random Brownian motion.

Settling of particulates is important since it offers a natural way for the removal of particulates from the air. Settling chambers are devices commonly used for dust collection. Nonetheless, it should be noted that such devices are only effective for large particles. Their efficiency drops considerably as the size of the particles to be collected decreases.

Particulate matter is subject to other physical phenomena besides just settling. For instance, they tend to scatter incident solar radiation and thus tend to decrease visibility. Such scattering behavior is greatly influenced by the size of the particulates. Particles smaller than 0.1 microns in diameter are comparable in size to the wave length of incident radiation and closely obey the laws of molecular scattering. On the other hand, particles much bigger than 1 micron in diameter are very much larger than the wave length of visible light and behave as macroscopic objects intercepting or scattering light in rough proportion to their cross-sectional area.[13]

Many interface phenomena can take place between particulates and the all-encompassing atmosphere. Such phenomena include adsorption, absorption, nucleation, and adhesion. Of these, nucleation is possibly the most important. It consists in the condensation of super-saturated vapor on the absorbed layer of a particle, thus forming a droplet or ice crystal. This is a most important mechanism in the process of rain formation.

Aside from the above-mentioned physical properties of particulates it is important to be aware of some of their more general chemical characteristics. It is difficult to make general statements about the chemical properties of particulate matter. There are many different types of particulates, and since each has a different characteristic it reacts and behaves differently. Nevertheless, some reaction systems such as gas-particle and particle-particle systems have been studied, especially for the size range

TABLE 7.4.

Nationwide particulate emissions (1968).[11]

Source	*Emissions, 10^6 tons/year*		*Percent of total*	
Transportation	1.2		4.3	
Motor vehicles		0.8		2.8
Aircraft		nil		nil
Railroads		0.2		0.7
Vessels		0.1		0.4
Non-highway use of motor fuels		0.1		0.4
Fuel combustion in stationary sources	8.9		31.4	
Coal		8.2		29.0
Fuel oil		0.3		1.0
Natural gas		0.2		0.7
Wood		0.2		0.7
Industrial processes	7.5		26.5	
Solid waste disposal	1.1		3.9	
Forest fires	6.7		23.7	
Structural fires	0.1		0.4	
Coal refuse burning	0.4		1.4	
Agricultural burning	2.4		8.4	
Total	28.3		100.0%	

around 0.1 micron where particle collisions are frequent. For instance, the reaction between ammonia gas and sulfuric acid mist was investigated by Robbins and Cadle.[14] Goetz and Pueschel[15] presented a study of a simplified model of basic photochemical aerosol formation in the air of Los Angeles, California.

From a chemical standpoint particulates can be divided into two groups, namely, organic and inorganic. Organic particles are mostly volatile compounds such as saturated and unsaturated aliphatic and aromatic hydrocarbons together with a variety of their oxygenated and halogenated derivatives, products of fossil fuel combustion, and motor-vehicle emissions. Most of the organic matter remains in the atmosphere as liquid aerosol when not reacting chemically or photochemically with gases or other particles.

The chemical nature of inorganic particles depends to a large degree on the emitting sources. For example, emissions from incinerators in Los Angeles County indicated 5–15% of the emissions' condensate to be sulfuric acid and the remaining 85–95% to be a mixture of silicon, lead, aluminum, calcium, iron, and some trace elements.[16]

Lead has been one of the particulates that has received the most attention in the past few years. The fight for the removal of lead from gasoline, however, was mostly based on lots of words and little technical expertise. As of this date, it cannot really be concluded that lead is a major atmospheric pollutant.* High concentrations of lead can of course be harmful and even fatal to man, however, most of the lead being emitted from pipe exhausts settles immediately on or near the roadway. Very little lead remains suspended in the atmosphere. This case emphasizes the need for increasing the awareness of the technical aspects of pollution in the part of managers, government officials, and individuals.

Sources of Particulate Emissions

Particulate emissions in the United States for the year 1968 are listed in Table 7.4. This table shows that of the 28.3 million tons of particulates emitted during 1968, 17.6 million tons came from fuel combustion, transportation, and industrial sources and 6.7 million tons came from natural sources such as forest fires.

* An airborne lead study was released in 1971 by the National Research Council on behalf of the Environmental Protection Agency. This study confirms the statements above. The NRC report states that for an average city resident exposure to lead levels produces a blood lead concentration "about half that necessary to cause biochemical changes in the body and one-fourth the level at which symptoms of lead poisoning begin to occur." Research by numerous other health authorities confirm that lead from auto-exhaust is not as serious an air pollutant as most people think.

7.5 GASEOUS POLLUTANTS

The emission of gaseous pollutants represents on the average more than 80 percent of the total atmospheric emissions in the United States. Comparing the psychological effects on man of both gaseous and particulate emissions, the latter has a more dramatic effect. Man is more concerned about the black smoke or dusty emissions that he can observe coming out of stacks than about (mostly) invisible gaseous pollutants. However, some such gases even in concentrations as small as a few tenths of parts per million (ppm) can cause damage to health and property not only as primary pollutants but also as products of complex photochemical reactions that take place in a polluted atmosphere.

For purposes of our discussion we will consider the following major gaseous pollutants:

(a) sulfur dioxide,
(b) nitrogen oxides,
(c) ozone,
(d) carbon monoxide, and
(e) hydrocarbons.

Sulfur Dioxide

Of all the sulfur compounds emitted to the atmosphere, sulfur dioxide is the most important air pollutant. Huge amounts of SO_2 (sulfur dioxide) are emitted each year as a product of solid and liquid fossil fuel combustion processes. The burning of such fuels constitutes the major source of sulfur dioxide pollution in the United States,[17] as well as in most industrialized nations.

Sulfur dioxide is a colorless, non-explosive gas under normal conditions. It is easy to detect by taste at concentrations as low as 0.3 ppm, but above 3 ppm it has a pungent and very irritating odor. SO_2 (sulfur dioxide) is highly soluble in water (113 g/1 at 20°C) and has a molecular weight of 64.06 and a density of 2.927 g/1 (of gas) at standard temperature and pressure.

Sulfur dioxide pollution of the air is the result of coal burning and combustion of petroleum products (Table 7.5). Metallurgical operations such as smelting (mostly of copper and zinc) contribute substantial amounts of sulfur dioxide to the atmosphere. Small amounts of sulfur trioxide, hydrogen sulfide and mercaptans form the remaining bulk of sulfur pollutants. In addition, as a result of the oxidation of atmospheric sulfur dioxide, some sulfuric acid and sulfate salts are formed. These normally

TABLE 7.5.

Nationwide emissions of sulfur dioxide from stationary sources (1966).*

Source	*Emissions, 10^6 tons/year*	*Percent of total*
Burning of coal		
Power generation	11.9	41.6
Other combustion	4.7	16.6
Combustion of petroleum products		
Residual oil	4.4	15.3
Other products	1.2	4.3
Smelting of ores	3.5	12.2
Refinery operations	1.6	5.5
Processing of coke	0.5	1.8
Manufacture of H_2SO_4	0.6	1.9
Miscellaneous	0.2	0.8
Total	28.6	100.0%

* Data modified from "Air Quality Criteria for Sulfur Oxides," (1969) U.S. Department of Health, Education and Welfare, National Air Pollution Control Administration Publication No. AP-50, Washington, D.C.

account for 5–20% of the total concentration of suspended *particulate matter* in urban air.[17]

Sulfuric acid can result from the oxidation of atmospheric sulfur dioxide[18] especially if a catalyst (manganese sulfate or some metal oxides) is present or it can result from a photochemical oxidation reaction.[19] The oxidation of sulfur dioxide in air may also occur when SO_2 is adsorbed on the surface of atmospheric aerosols. The reaction produces sulfate salts or sulfuric acid.

Junge and Ryan[20] studied the catalytic oxidation of atmospheric sulfur dioxide in order to be able to infer a sensible mechanism to account for the observed concentrations of the sulfate ion (SO_4^{--}) in rain, fog, and smog. Their conclusion was that the catalytic oxidation of SO_2 in the presence of NH_3 (ammonia) can explain the mean features of SO_4^{--} formation in the atmosphere. It appears that this oxidation reaction is not photosensitive during normal daylight conditions. The same authors did some experiments in which the influence of pH on the catalytic oxidation of SO_2 was measured. They found that the formation of SO_4^{--} stopped around a pH of 2.00. If ammonia was present in the air or in the catalyst, oxidation of sulfur dioxide continued since the ammonia would neutralize the acid as it was formed.

Sulfur dioxide in air can also undergo a slow photochemical oxidation under the action of sunlight in the presence of both oxygen and moisture. The resulting product, as in the catalytic case discussed above, is sulfuric acid. Gerhard and Johnstone,[19] and Hall[21] have extensively studied this important photochemical process. Their results can be summed as follows:

(a) The photochemical oxidation of SO_2 is a first order reacticn with respect to SO_2 concentration;
(b) The reaction rate is slow, amounting to about 0.1 to 0.2% conversion per hour under intense natural sunlight; and
(c) The rate seems to be unaffected by humidity, the presence or absence of salt nuclei, or the concentration of nitrogen dioxide in the range of concentrations they studied.

Dainton and Iving[22] studied the reaction between SO_2 and hydrocarbons that are emitted into the atmosphere mostly as part of automotive exhaust emissions. They introduced the SO_2 and hydrocarbons into a cylindrical quartz cell illuminated by a hot mercury arc. They were able to observe the formation of a mist in the reaction vessel which after settling formed a colorless or pale-yellow non-volatile oil.

Schuck, Ford, and Stephens[23] found that the photochemical oxidation of SO_2 mixed with olefins and nitrogen dioxide leads to the formation of an aerosol. The major product of this reaction is sulfuric acid, which being hygroscopic tends to absorb water to form droplets of sulfuric acid.

Sulfur dioxide can be dangerous to health, especially in concentrations above 5 ppm. Its effect is mostly an irritation of the upper respiratory tract, however, workers normally exposed to high concentrations of SO_2 seem to build an immunity to this gas and in many cases show no irritation in their respiratory tract. More damaging effects are observed when subjects are exposed to a combination of SO_2 and sulfuric acid mist in the air.

Oxides of Nitrogen

The most common oxides of nitrogen in a polluted atmosphere are nitric oxide (NO), nitrogen dioxide (NO_2), and some nitric and nitrous acids and their salts.

Fuel combustion in stationary sources is the prime contributor of these atmospheric pollutants (Table 7.6). Under typical combustion conditions nitric oxide is formed by the direct combination of molecular nitrogen and oxygen. A high temperature is needed for this process to take place due to the extremely stable nature of the N–N bond.[24] Once NO is in the atmos-

TABLE 7.6.

Nationwide emissions of oxides of nitrogen (1968).[11]

Source	*Emissions, 10^6 tons/year*		*Percent of total*	
Transportation	8.1		39.3	
Motor vehicles		7.2		34.9
Aircraft		nil		nil
Railroads		0.4		1.9
Vessels		0.2		1.0
Non-highway use of motor fuels		0.3		1.5
Fuel combustion in stationary sources	10.0		48.5	
Coal		4.0		19.4
Fuel oil		1.0		4.8
Natural gas, LPG, and kerosene		4.8		23.3
Wood		0.2		1.0
Industrial proc.	0.2		1.0	
Solid waste disp.	0.6		2.9	
Miscellaneous	1.7		8.3	
Forest fires		1.2		5.8
Coal refuse and agricultural burning		0.5		2.5
Total	20.6		100.0%	

phere it is oxidized and forms NO_2. This reaction is especially fast if sunlight is available. The oxidation process is written:

$$2NO + O_2 \rightarrow 2NO_2 \tag{7.5-1}$$

Nitrogen dioxide is the most efficient absorber of the sun's light that reaches the earth's surface. It absorbs virtually the whole visible and ultraviolet spectrum. This capability is responsible for the occurrence of a series of photochemical reactions. First of all, the absorption of ultraviolet light below 3700 Å photolyzes the NO_2 molecule into NO and O (atomic oxygen) in a very rapid reaction, that is,

$$NO_2 + \text{energy} \rightarrow NO + O \tag{7.5-2}$$

The atomic oxygen formed by (7.5-2) immediately reacts with molecular oxygen to yield ozone (O_3), that is,

$$O + O_2 + M \rightarrow O_3 + M \tag{7.5-3}$$

In the above reaction, M is a third body which accounts for any excess in the amount of energy transferred. M in the atmosphere can be molecular nitrogen (N_2) or molecular oxygen (O_2).

The final reaction of the process involves the NO and O_3 formed by (7.5-2) and (7.5-3) respectively. The result is

$$NO + O_3 \rightarrow NO_2 + O_2 \tag{7.5-4}$$

All the above reactions, however, are further complicated by the presence of hydrocarbons in the atmosphere.

Many laboratory experiments have been carried out to elucidate the mechanisms of the atmospheric photochemical reactions of nitrogen oxides with hydrocarbons. These experiments have mainly involved the irradiation of mixtures of nitrogen oxides and olefins, and irradiation of simulated and authentic auto-exhaust gases. For instance, Stephens[25] studied the reaction between olefins and nitrogen oxides in the atmosphere. He found that some of the resulting products were carbonyl compounds, alkyl nitrate, peroxyacetyl nitrate (PAN) and peroxypropionyl nitrate (PPN). Some of these compounds are irritating to the eyes while others are toxic to plant life.

Altshuller et al[26] irradiated the systems propylene-nitrogen oxide and propylene-nitrogen oxide-sulfur dioxide. The reaction products were: formaldehyde, acetaldehyde, PAN, methyl nitrate, carbon monoxide, ozone, and various light-scattering aerosols. All these products are smog-forming.

Tuesday[27] investigated the atmospheric photochemical oxidation of trans-butene-2 and nitric oxide using a long-path infrared chamber coupled to an infrared spectrophotometer. The products that he obtained after irradiation were acetaldehyde, formaldehyde, PAN, methyl nitrate, ozone, carbon monoxide, carbon dioxide, water, methyl alcohol, formic acid, and some intermediate products. He reported that as soon as the irradiation began acetaldehyde, methyl nitrate, and carbon dioxide were formed, however, PAN was not formed until the oxidation of nitric oxide was complete.

Pomonowsky et al[28] have confirmed the eye-irritant properties of the oxides of nitrogen, especially in the presence of hydrocarbons. They measured this irritation using a selected panel of human subjects under simulated conditions of urban vehicle-emission levels.

Ozone

Ozone is a colorless gas with a very characteristic odor. It strongly absorbs sunlight, especially in the 2000–3000 Å range. This property of ozone is of utmost importance in helping to preserve the atmospheric heat budget. In addition, ozone serves as a screen preventing the intense ultraviolet solar radiation from reaching us.

The high concentrations of ozone observed in a polluted atmosphere are mostly the result of the interaction of molecular and atomic oxygen as described by equation (7.5-3). In addition, as discussed in talking about the oxides of nitrogen, some ozone is formed from the photochemical oxidation of many olefins, paraffins, and alkyl-benzenes[29] in the presence of nitric oxide or nitrogen dioxide. Stratospheric ozone is of no consequence from a pollution standpoint since it has been proved by McKee[30] that stratospheric ozone does not reach ground level.

As pointed out in the preceding discussion of the oxides of nitrogen, many laboratory experiments have been carried out to study the reactions of ozone with nitric oxide, as well as with many olefins and other hydrocarbons. These studies have shown the complexity of such reactions.

While many laboratory experiments have been carried out to understand the reactions between ozone and other atmospheric constituents, it should be noted that the concentration of these constituents in the experiments have usually been higher than those observed in photochemical smog. Therefore, one has reasons to suspect that products formed under laboratory conditions may not be formed in polluted air.

Carbon Monoxide

Carbon monoxide (CO) arises in enormous amounts from the incomplete combustion of carbonaceous fuels. It is the most widely distributed and the most commonly occurring air pollutant. Inhalation of carbon monoxide has produced poisoning to man since he began to use coal for domestic and heating uses.

Carbon monoxide from man-made activities is emitted by a variety of sources that can be grouped into four main categories (Table 7.7), namely, transportation, stationary sources of fuel combustion, industrial processes, and solid waste disposal. Table 7.7 summarizes these categories as they apply to the United States. From this table one can see that the biggest offender in terms of carbon monoxide is the automobile.

Cars built prior to the model year 1968 have no emission-control devices for CO and produce average tailpipe emissions of 73 grams of CO/mile. Exhaust standards set by the U.S. government allow a 1971 automobile to emit from the tailpipe no more than 23 grams of CO/mile traveled. Emission standards prescribed by the Clean Act[31] for 1975–76 cars require a 90% reduction in carbon monoxide emissions as compared with 1970–71 cars. Similar reductions will go in effect with respect to hydrocarbons, and oxides of nitrogen.

TABLE 7.7.

Nationwide emissions of carbon monoxide (1968).[11]

Source	*Emissions 10⁶ tons/yr*		*Percent of total*	
Transportation	63.8		63.8	
Motor vehicles		59.2		59.2
Gasoline		59.0		59.0
Diesel		0.2		0.2
Aircraft		2.4		2.4
Railroads		0.1		0.1
Vessels		0.3		0.3
Non-highway use of motor fuels		1.8		1.8
Fuel combustion in stationary sources	1.9		1.9	
Coal		0.8		0.8
Fuel oil		0.1		0.1
Natural gas		nil		nil
Wood		1.0		1.0
Industrial processes	9.7		9.6	
Solid waste disposal	7.8		7.8	
Miscellaneous	16.9		16.9	
Forest fires		7.2		7.2
Structural fires		0.2		0.2
Coal refuse burning		1.2		1.2
Agricultural burning		8.3		8.3
Total	100.1		100.1	

In spite of the fact that every year millions of tons of CO are pouring into the atmosphere, the background levels of CO are not rising significantly at the present time. Pressman et al[32] published the results of their theoretical and laboratory investigations of natural chemical processes by which CO is oxidized and removed from the atmosphere. They discuss in their paper the photochemical oxidation of CO into CO_2 in the stratosphere and conclude that this is a significant sink for carbon monoxide. They also discuss the oxidation of CO by ozone and conclude that this reaction is important as far as CO depletion is concerned if it is catalyzed by atmospheric methane.

The biosphere provides another removal process for CO through a variety of plants and microorganisms that are able to metabolize CO.[33]

The ocean is not a sink for carbon monoxide. On the other hand, the result of measurements by Swinnerton et al[34] conclusively show that the ocean is a source rather than a sink for atmospheric CO.

Regardless of which of the above mechanisms prevails Robbins et al[35] have estimated that the mean residence time of CO in the atmosphere is about five years.

Hydrocarbons

Hydrocarbons are organic compounds consisting primarily of carbon and hydrogen. Most of the hydrocarbons found in polluted atmospheres are in the gaseous phase.

Hydrocarbons are classified in three categories, namely, acyclic, alicyclic, and aromatic. The first category corresponds to hydrocarbons with a linear chain of carbon atoms with or without branching chains. The aromatic group is that whose atoms are arranged in benzene rings with six atoms of carbon and only one atom of hydrogen attached to each atom of carbon. The alicyclic category corresponds to all those hydrocarbons having a ring other than benzene.

Hydrocarbon pollutants in the atmosphere are primarily the result of the incomplete combustion of fossil fuels. The major source of hydrocarbon emissions in urban communities is the automobile. Forest fires, organic solvent evaporation, and agricultural burning also contribute substantial amounts of hydrocarbons to the environment.[36] Table 7.8 lists hydrocarbon emissions in the United States during 1968.

A substantial reduction of hydrocarbon emissions is expected in 1975–76 when, as previously pointed out, cars would be required to reduce by 90% hydrocarbon emissions as compared with 1970–71 cars. Research is presently underway both in Detroit and in oil-company laboratories to achieve the required goals.

In previous parts of this section we have discussed some of the reactions where hydrocarbons intervene. Laboratory experiments have shown that the most important reactions converting hydrocarbons and other organic substances into photochemical products are those involving atomic oxygen, ozone, and certain free radicals. In the absence of reactive species such as ozone, atomic oxygen, or nitrogen dioxide, hydrocarbons are quite inert pollutants. Some laboratory experiments involving hydrocarbons have already been discussed in connection with our preceding discussions of sulfur dioxide, ozone, and oxides of nitrogen. Other experiments are described in more advanced works.[37–40]

Aldehydes are the major products formed by the photochemical oxidation of hydrocarbons and the reaction of hydrocarbons with ozone, atomic oxygen, or free radicals. Aldehydes are considered as some of the most

TABLE 7.8.

Nationwide hydrocarbon emissions (1968).[11]

Source	*Emissions, 10^6 tons/year*	*Percent of total*
Motor vehicles	15.6	48.8
Aircraft	0.3	0.9
Railroads	0.3	0.9
Vessels	0.1	0.3
Non-highway use of motor fuels	0.3	1.0
Coal	0.2	0.6
Fuel oil	0.1	0.3
Natural gas	nil	nil
Wood	0.4	1.3
Industrial processes	4.6	14.4
Solid waste disposal	1.6	5.0
Forest fires	2.2	6.9
Structural fires	0.1	0.3
Coal refuse burning	0.2	0.6
Agricultural burning	1.7	5.3
Organic solvent evaporation	3.1	9.7
Gasoline marketing	1.2	3.7
Total	32.0	100.0%

important smog-causing agents. They are dissociated by sunlight to yield alkyl radicals, formyl radicals, and hydrogen atoms. In addition, these radicals as well as hydrogen will react with oxygen atoms to form the corresponding peroxyl radicals.

7.6 PHOTOCHEMICAL SMOG

The chemical composition of polluted air in a given community will depend on the chemical nature of the pollutants emitted by the many different sources. Air masses, moreover, are dynamic systems in terms of meteorological parameters and in terms of the chemical reactions that take place between different substances in the air. In addition, to all these factors, many substances under the influence of solar energy, undergo several complex reactions known as photochemical processes. The energy range at which atmospheric photochemical reactions occur lies in the 2900 Å to 7000 Å region of the solar spectrum, however, the region of most photochemical importance is the portion of the ultraviolet range between 2900 Å and 4000 Å. The 2900 Å lower limit is set by the cutoff of the radiation due to stratospheric ozone.

During the early forties Los Angeles began to suffer a type of pollution nowadays termed "photochemical smog." This pollution results from a series of photochemical processes that take place between organic compounds in the atmosphere and oxides of nitrogen. Hydrocarbons and aldehydes are two of the most important organic contaminants responsible for the production of "photochemical smog."

In very general terms the formation of smog can be described as follows. As a result of the interaction of a photon of sunlight with a molecule in a primary photochemical process, a stable molecule may be produced directly or some unstable product may be formed which may undergo secondary reactions. The primary photochemical products consist of radicals or free atoms that will take part in secondary reactions which may lead to stable molecules. These in turn may or may not participate in other different reactions. Some of the end products of these reactions are irritating to the eye and/or to the respiratory system of some people.

Several factors are responsible for the occurrence of smog in a given area such as Los Angeles. These can be summarized as follows:

(a) the existence of atmospheric stability in the area,
(b) a high degree of urban development creating a poorly ventilated environment,
(c) a high consumption of fossil fuels, and
(d) a steady sunshine to trigger the photochemical process.

Photochemical smog is no longer a phenomenon seen only in Los Angeles. Many urban areas in the United States and abroad where the four factors listed above exist also suffer a smog problem. The effects of this smog will be discussed in a later section of this chapter.

In concluding, it should be noted that the term smog is improperly used, since it really refers to a combination of smoke and fog and this is incorrectly used to describe what should be termed "photochemical pollution."

7.7 METEOROLOGICAL EFFECTS ON AIR POLLUTION

It is well-known today that meteorological conditions play a very important role in controlling the extent of air pollution in a community. The concentration of pollutants varies according to some non-linear relationship between strength of sources of pollution and meteorological parameters, such as, wind speed, vertical gradient of temperature, variations in rainfall, etc. The ways in which atmospheric dilution processes, such as, mixing and diffusion take place are very complex. The motion of turbulent diffusion and multiple-source urban diffusion are so complicated mathematically that they are beyond the scope of this chapter. Readers may refer to some excellent books and papers written on these specific subjects.[41–44] In this section, we shall concentrate on describing the most important features of the meteorological parameters that influence the dilution of air pollution.

Atmospheric Stability and Instability

One of the fundamental concepts of air pollution meteorology is atmospheric stability. This concept can be explained in terms of the capacity of small parcels of air to move vertically through a given layer of the atmosphere. A layer is said to be stable when it withstands vertical air motion. A layer is unstable when it helps the vertical movement of air masses.

Air behaves much like an ideal gas. Changes in temperature produce changes in volume and pressure according to well-known gas laws related to thermodynamical properties involving heat and work.

When a parcel of air moves vertically through a free atmosphere, changes in pressure and volume will occur while the external pressure of the surrounding air either increases or decreases. If the parcel rises, volume will increase. These changes will be associated with temperature variations. An expanded parcel will cool and a compressed one will heat-up, as they move respectively upward or downward. If temperature changes occur as a consequence of compression or expansion (i.e. heat is not gained nor lost from the parcel) the process is said to be adiabatic.

Under normal daytime conditions the earth's surface warms up more readily than the air above and the normal lapse rate (temperature gradient) becomes slightly larger than the temperature decrease of the adiabatic expansion of any air-rising parcel. Under this situation, the lapse rate then becomes superadiabatic. As the air parcel ascends it will cool adiabatically, but if the superadiabatic condition exists the parcel of air will be warmer than the surroundings and will thus continue to rise. Similarly, if the parcel of air is lowered in altitude in a superadiabatic surrounding atmosphere, the parcel will be denser than ambient and will continue to sink. Thus, it is clear that a superadiabatic condition favors vertical motions and therefore the atmosphere becomes unstable.

Under certain conditions, the normal lapse rate is reversed and temperature increases with altitude. This condition usually occurs after sunset when the earth's surface begins to emit blackbody radiation at a rate faster than its gaseous surroundings. As a parcel of air ascends into the atmosphere it becomes cooler and consequently more dense than the surrounding air. Therefore, the upward motion is held back and the parcel may descend to its original position. In the same way, a parcel sinking through a layer is itself warmer (less dense) than its surroundings and thus the descending motion is held back and the parcel may rise back to its original position. These acts result in a very stable condition called inversion.

Atmospheric Diffusion

The main concern of air pollution meteorology is to estimate the probability of having a given pollution concentration in a determined area, considering the particular set of meteorological and topographic conditions of such area. In addition, important goals are to evaluate the diffusion of pollutants from a point-emitting source, and to establish mathematical models for urban multiple sources.[41] Several meteorologists have worked out simplified models to study the diffusion of stack effluents, even though these models do not give exact real solutions due mainly to the large number of parameters involved. Fortunately, in many cases a large number of parameters are negligible and these models can be applied to a great variety of practical objectives. Therefore, it is possible to obtain acceptable results, as in the case of estimating possible hazards from some stack effluents in operating plants or in designing new factories, or even in the case of radioactive material release as a result of an accident from an atomic power plant. According to particular circumstances and meteorological data available, the estimating procedures may be varied in each particular case.

Plume Types

Stack effluents are called *plumes*. A visible plume of smoke takes on a variety of forms depending on the weather conditions and time of day. For instance, for cloudy weather in open country and under moderate wind speeds, the smoke forms a fairly straight, well-defined trail which increases steadily in width and height as the distance from the source increases. If there is sufficient solar heating of the ground surface, the appearance of the trail is much more irregular. On the other hand, in clear nights there is appreciable cooling of the ground and if the wind is light the vertical-spread and rise of the smoke trail downwind forms a compact plume visible from far away.

Church[45] has divided the dispersion behavior of stack effluents into five classes based on ambient atmospheric stability. These types of plumes are depicted in Figure 7.3, and can be described as follows:

1. Looping plume. Occurs with strong superadiabatic lapse and high degree of turbulence. The daytime occurrence is favored by a clear sky, intense solar heating, and unstable thermal conditions.

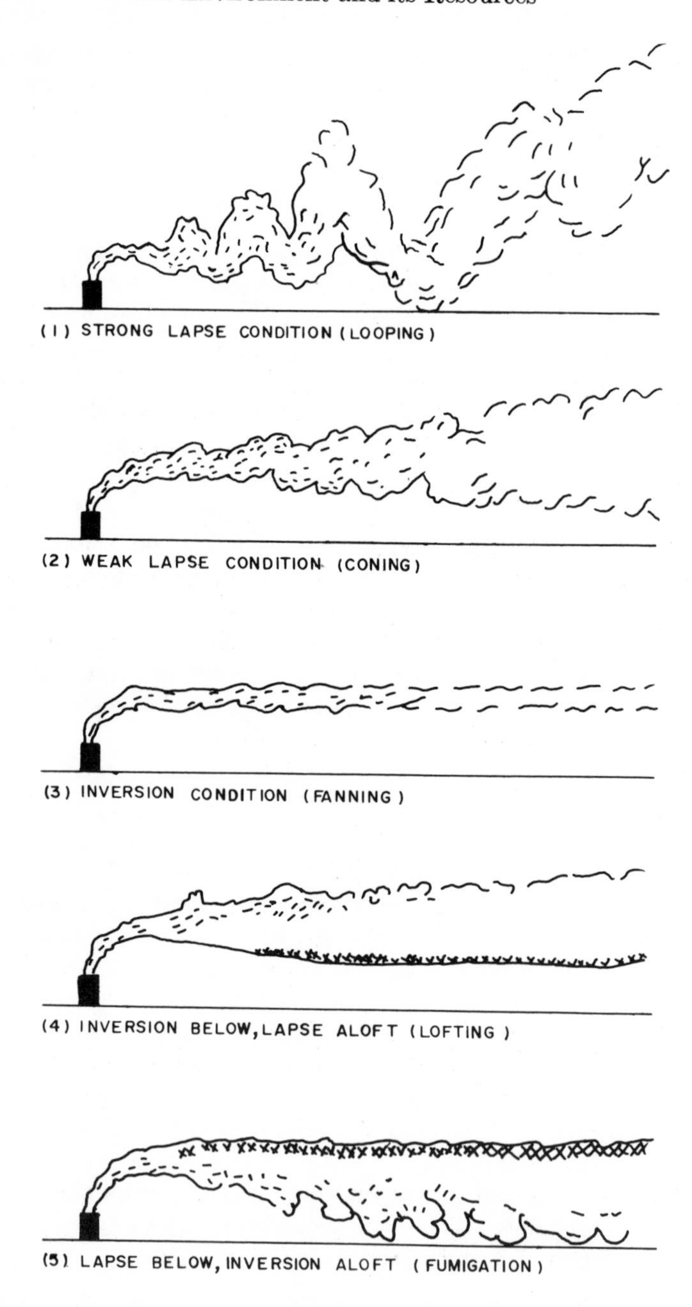

Figure 7.3. Schematic representation of stack gas behavior under various atmospheric conditions.

2. Coning plume. Occurs under weak lapse and nearly thermal conditions. Cloudy skies favor this type of plume because clouds reduce thermal effects.
3. Fanning plume. Occurs under inverted or nearly isothermal lapse rate, and stable conditions. Mostly occurs at night and it is favored by light winds and clear skies when the earth's surface is cooled by outgoing radiation.
4. Lofting plume. It is caused by a stability condition just below the stack exit coupled with a temperature inversion. The unstable layer is above the plume and larger diffusion goes upward. It occurs mostly near sunset on fair days with moderate winds and when turbulence is found in the exit layer. Light winds and little or no turbulence occur in the layer below the plume. This type of plume favors low ground level concentration.
5. Fumigation plume. This plume is the opposite of lofting. It is caused by unstable air below the plume exit and stable air above it. It occurs when nocturnal inversion has been broken by the morning sun which warms up the ground surface and thus a stable layer develops from the ground upwards.

Stack Effluent (Plume) Diffusion

The mathematical methods for obtaining concentrations of stack effluents at ground level have been formulated by some researchers. Some of these methods give the equations for determining the height of the plume above the ground surface, or effective stack height. Other methods give the formulation for diffusion of stack or smoke about the plume axis and the deposition of particles at a given distance from the foot of the stack.

Fick's differential equation of molecular diffusion of gases has been the starting point of most mathematical theories of diffusion from emitting sources. Fick's law states that "diffusion of material is in the direction of decreasing concentration and is proportional to the concentration gradient."

Fick's equation for diffusion in a turbulent wind with a given concentration of suspended matter (mass/unit volume) is

$$\frac{d\chi}{dt} = \frac{\partial}{\partial x}\left(K_x \frac{\partial \chi}{\partial x}\right) + \frac{\partial}{\partial y}\left(K_y \frac{\partial \chi}{\partial y}\right) + \frac{\partial}{\partial z}\left(K_z \frac{\partial \chi}{\partial z}\right) \qquad (7.7\text{-}1)$$

where χ = the concentration at a given point in the plume (mass/volume), and K_x, K_y, and K_z are the diffusion coefficients in the x, y, and z directions respectively.

Sutton[46] has proposed some approximate solutions of equation (7.7-1).

These have been widely applied for diffusion near the ground surface. One of these equations corresponds to the solution for a continuous point source based on the statistical properties of turbulence. This equation is written:

$$\chi(x, y, z) = \frac{2Q}{\pi K_y K_z \bar{u} x^{2-n}} \exp\left[-x^{n-2}\left(\frac{y^2}{K_y^2} + \frac{h^2}{K_z^2}\right)\right] \qquad (7.7\text{–}2)$$

where

χ = downwind ground level concentration at a point (x, y, z); (mass/volume),
Q = source strength; (mass/time),
K_y, K_z = diffusion coefficients in the y and z planes, respectively,
n = turbulence index (nil-dimensional),
$\bar{u}$ = mean wind speed; (length/time), and
h = source height.

The turbulence parameter n approaches zero in a very turbulent air and increases toward a maximum value of unity in a stable atmosphere. The values of K_y and K_z are dependent on the mean wind velocity and the degree of gustiness. Gustiness is defined as the series of irregularities in the velocity of the wind caused either by the mechanical effect of surface irregularities or by convection currents due to surface heating.[47]

Sutton[46] has also worked out an equation for finding smoke or gas concentration at any location downwind of a point-elevated source. This equation is expressed as:

$$\chi = \frac{Q \exp[-y^2/(x^{2-n}K_y)]}{\pi K_y K_z \bar{u} x^{2-n}} \left\{\exp\left[-\frac{(z-h)^2}{K_z^2 x^{2-n}}\right] + \exp\left[\frac{(z-h)^2}{K_z^2 x^{2-n}}\right]\right\} \qquad (7.7\text{–}3)$$

The maximum concentration at ground level can be obtained from the above equation and it is found to be directly proportional to source strength and inversely proportional to the wind speed and the square of the plume height.

Pasquill[48] has devised a mathematical method for computing the diffusion of plumes from elevated or ground-level sources in terms of the height and width of the plume of airborne material. Even though the basis of his model is quite different from Sutton's the resulting equations are in excellent agreement.

In addition to the works by Sutton and Pasquill, other authors have carried out research to relate atmospheric dilution of pollutants to meteorological conditions. The readers can refer to more specialized papers for this information.[49]

In conclusion, it should be emphasized that meteorological conditions control the dilution of all pollutants dumped by man and nature into the atmosphere. If conditions are right smog will result. Similarly, under different conditions the same amount of pollutants could easily diffuse and be harmlessly lost.

7.8 SOURCES OF ATMOSPHERIC POLLUTION

All man-made air pollution can be said to come from one of four basic sources, namely, mobile sources, stationary sources, solid waste disposal, and miscellaneous sources. The group of mobile sources covers all modes of transportation such as cars, trucks, buses, planes, railroads, etc.

Stationary sources include power plants, industrial processes, steel mills, oil refineries, smelters, etc. Solid waste disposal can be classified as a stationary source, however, it is convenient to consider it separately because of the different nature of emission products. Solid waste disposal includes emissions from municipal and domestic incinerators, sewage treatment plants, and collection and disposal of refuse. Under the category of miscellaneous we include fires, coal refuse burning, agricultural burning, etc.

Mobile Sources

Mobile sources are the major source of carbon monoxide and hydrocarbon pollution in the United States. They also rank second in nitrogen oxides' emissions (see Table 7.3). Of the 63.9 million tons of carbon monoxide poured into the air during 1968 from mobile sources, 59 million were contributed by gasoline-powered vehicles.[11] Of course, as pointed out in section 7.5 a great portion of the carbon monoxide poured into the atmosphere is lost through various sinks.

Hydrocarbons are also emitted in large quantities from mobile sources. These emissions originate in four areas of a gasoline-powered vehicle, namely, the carburetor, fuel tank, crankcase, and exhaust. The latter is by far the major culprit of the four.

Around fifty individual hydrocarbons have been identified in exhaust gases in varying amounts. These are produced from all fuels under all modes of engine operation.[5] Some of the hydrocarbons identified are acetylene, ethylene, isobutane, propadiene, benzene, and toluene.

Polynuclear aromatics and long-chain aliphatic hydrocarbons have also been isolated from automobile exhaust. Of these we can mention anthracene, pyrene, perylene, benzo-(α)-pyrene, benzo-(β)-pyrene, and benzo-(κ)-fluoranthene. The last three compounds are reported to be carcinogenic, but of these, benzo(α) pyrene is the most predominant. All three compounds, however, are present in very minute quantities.

Stephens and Burleson have studied the distribution of light hydrocarbons in ambient air.[50] They have reported that the composition of their samples resembled that of auto-exhaust gases plus natural gas and gasoline vapor. Neligan[51] in a study of hydrocarbons in the Los Angeles atmosphere detected fifty seven different hydrocarbons in samples of urban air.

Lead content in the atmosphere has increased during the last two decades.[52] The most important source of lead is the exhaust of automobiles. The emitted lead comes from the gasoline antiknock additives. Use of low-lead fuels will help a long way to reduce these emissions.

Mobile sources also contribute roughly forty percent of nitrogen oxides' emissions, and four percent of total particulate emissions.

On the plus side of the emissions' picture we should note that between 1966 and 1968 there was a one-million tons/year decrease in hydrocarbon emissions from mobile sources.[11] Controls presently planned for mobile sources should reduce hydrocarbon emissions to much lower levels.

A problem of utmost importance in reducing emissions from mobile sources, mostly cars, is a human one. Most anti-pollution devices tend to decrease the performance of an engine. Thus, some people find it rather tempting to disconnect or disassemble them. In addition, many people do not give their cars proper care and emissions in turn tend to increase. To solve these problems individual state governments or the federal government may need to develop strict periodic inspection procedures covering anti-pollution devices.

Stationary Sources

Stationary sources of air pollution are most abundant in urban, industrialized areas. The most important characteristic of these sources is their variability. The pollutants emitted from a cement plant, for instance, are quite different from those emitted by a copper smelter. The abatement procedures to be used in each case are also drastically different.

Consumption of large amounts of heavy fuel oil and coal at factories, power plants, and for domestic heating constitute the major source of sulfur oxides' emissions in the United States. These fuels contain most of the time between 0.5 to 4.0 percent sulfur. Coal burning is the biggest culprit as far as SO_2 emissions are concerned (Table 7.5) contributing over half of the total emissions of this pollutant in the United States.

Stationary sources also contribute to the atmosphere of our industrial cities a large amount of particulate matter. The type of particles emitted varies depending on the industry, chemical process, or class of fuel consumed.

It is estimated that the iron and steel industry produced nearly 25% of the total particulate emissions by industry in the United States during 1971.[7] The cement industry contributes a large amount of dust to the atmosphere in spite of highly efficient collection devices. Dust-fall rates of 25 tons per square mile have been reported even in areas adjacent to efficiently controlled kilns.[53] During 1971 cement-plant emissions in the United States have been estimated to be 0.51 million tons while lime plants contribute 242,000 tons of particulates.[7]

In kraft pulp mills particulate emissions originate from various phases of the processing operations. This industry's contribution to particulate emissions during 1971 in the United States has been estimated as around 100,000 tons. Emissions from kraft pulp mills have been extensively controlled by the use of wet scrubbers, cyclones, and electrostatic precipitators.

Petroleum refineries contribute some pollutants to the air environment especially in the form of dust, smoke, a variety of smog-forming hydrocarbons, and sulfur dioxide.

Most chemical operations contribute pollutants of one kind or another to the environment. The substances discharged vary depending on the type of process involved and the control devices installed.

Solid Waste Disposal

Refuse disposal as discussed in the preceding chapter is a major and growing concern of municipal authorities especially in urban areas. Every year millions of tons of garbage must be disposed of, producing as a by-product various air pollution problems. In many communities these air pollution problems are enhanced by the practice of open dump burning.

Refuse incineration contributes a large amount of the total fly-ash and soot emissions. Domestic incineration is still practiced in many communities in the United States where the lack of adequate refuse collection facilities and municipal incinerators forces the residents to backyard burning of any kind of debris.

The Bureau of Solid Waste Management of the United States Department of Health, Education, and Welfare recently undertook a survey on solid waste disposal practices.[54] Results of this survey show that the average collection rate of solid waste in the United States is about 38.5 pounds weekly per person. From results obtained during this survey, the government has been able to estimate the distribution of solid waste disposal practices in the United States (Table 7.9). Note from this table that during the period 1966 to 1968 municipal incineration and sanitary landfill increased while on-site incineration and open dumps decreased. This is a favorable trend that hopefully will continue at a more accelerated pace.

TABLE 7.9.

Solid waste disposal practices (1966, and 1968).[11]

Disposal method	*1966*	*1968*
	10^6 tons/year	
Municipal incineration	16	19
On-site incineration	57	55
Sanitary landfills	10	29
Open dumps	228	218
burned	77	82
non-burned	151	136
Wigwam burners	27	27
Hog feeding	1	1
Composting, treatment plants, etc.	19	18
Total	358	367

Miscellaneous Sources

In addition to the sources of pollution discussed above some air pollution occurs as the result of:

(a) forest fires,
(b) structural fires,
(c) coal refuse burning,
(d) organic solvent evaporation,
(e) gasoline marketing, and
(f) agricultural burning.

These miscellaneous sources are important since they contribute 17% of total carbon monoxide emissions, 34% of particulate emissions, 8% of emissions of NO_x, and 26.5% of total hydrocarbon emissions. Forest fires are particularly important from an air pollution standpoint since they contribute as much to hydrocarbon pollution as half of all industrial emissions, and they account for as much particulate pollution as all industrial processes in the United States.

7.9 MEASURING AIR POLLUTANTS

To adequately analyze the air pollution problem in a given area or in a particular industry, data about pollutants in the atmosphere is of utmost importance. Such data is best obtained by employing a continuous (automatic) air monitoring program. In the early sixties the Public Health Service developed the Continuous Air Monitoring Program (CAMP). Such stations were capable of measuring automatically the concentrations of carbon monoxide, sulfur dioxide, nitrogen dioxide, nitric oxide, total oxidants, and total hydrocarbons.[55] Many such stations have been installed throughout the entire United States, however, as many as 10,000 measuring stations may be needed to adequately cover the whole country.

A monitoring program in a community is a very important undertaking. A local government may initiate such a program by handling pollution complains as they are received and running spot checks on suspected sources of pollution.[56] Aerial surveys are also valuable in determining meteorological, topographic, and geological factors that may affect air pollution.

The pollutants measured during a monitoring program will vary depending on the purpose of the study. Among the most important parameters measured are:

1. reduction of visibility due to smoke,
2. particulate matter,
3. sulfate in atmospheric particulates,
4. nitrate in atmospheric particulates,
5. material deterioration,
6. total oxidants,
7. aliphatic aldehydes,
8. sulfur dioxide, and
9. NO_x (NO and NO_2).

Smoke can be measured manually using the Ringelmann Chart.[57] This chart measures the smoke being emitted from a plume against an opacity chart. There are also automatic samplers that are capable of measuring smoke levels at preset intervals.

Particulate emissions can be measured using various devices such as dustfall jars, electrostatic precipitators, light scattering devices, impaction and inertia collectors, AISI automatic tape samplers, and the Hi-volume apparatus. To obtain comparable results during a study one must always use the same technique for measuring particulate emissions.

Sulfate present in suspended atmospheric particulates can be measured using the turbidimetric barium sulfate ($BaSO_4$) method.[58,59] Nitrate in atmospheric suspended particulates can easily be determined by the 2,4 xylenol method.[59,60]

Deterioration of various materials when in contact with the atmosphere can be measured by exposing panels of such materials to ambient air.

Common materials whose deterioration is monitored are cotton cloth, nylon cloth, and steel.[61]

Oxidants, including ozone, are commonly measured by either the neutral-buffered potassium iodide (KI) method or by the alkaline potassium iodide method.[62] The presence of these pollutants is used to indicate the presence of photochemical reaction products.

Aliphatic aldehydes present in ambient air can be determined by the MBTH (3-methyl-2-benzothiazolone hydrazone hydrochloride) method.[59,63]

Sulfur dioxide can be manually measured by the West and Gaeke method[64] or the hydrogen peroxide method.[65] The former is best if the concentration of sulfur dioxide is between 0.005 and 5 parts per million (ppm), and if the concentrations of ozone and nitrogen dioxide are less than the concentration of sulfur dioxide. The hydrogen peroxide method requires simpler equipment and is easier to carry out than the West and Gaeke method. It can be used if the SO_2 concentration is between 0.01 and 10 ppm. Automatic methods for determining SO_2 include electroconductivity, potentiometric, photometric[66] and air ionization techniques.

Nitrogen dioxide can be determined manually by the Saltzman method.[59] This method is applicable if the NO_2 is present in the range from a few parts per billion to about 5 ppm. In addition, this method is applicable for determining the concentration of nitric oxide (NO) in the atmosphere. To accomplish this the NO is converted to an equivalent amount of NO_2 by passing the gas through a permanganate bubbler.[67]

To measure gaseous pollutants, as discussed above, one should begin by taking adequate samples. This can be done manually by using flasks or impingers. Suitable aqueous solutions, activated charcoal, or silica gel can be used as adsorbents. Air at the spot to be sampled is driven through these adsorbents by aspirators. Fully automated samplers are also available.

7.10 EFFECTS OF ATMOSPHERIC CONTAMINATION

Pollution of our air has brought about new health hazards to life on this planet. Such pollution affects man, animals, plant life, materials, and visibility. Many medical studies have been carried out to elucidate health hazards related to air pollution and to deal with the effects of pollution on groups of people living or working in a given area.

The effects of air pollution on man, animals, and plants may be classified into three main categories:

(a) Acute effects. These occur as a result of short exposures to comparatively high pollution levels.

(b) Chronic effects. These occur as a result of comparatively low levels of pollution over large periods of time.
(c) Long-term effects. These are the cumulative effects which occur after long-term coexistence of biological systems with air contaminants.

Effects on Humans

Air pollution affects population groups in different ways. Infants, the aged, those with preexisting chronic respiratory (emphysema and bronchitis) or cardiovascular diseases, heavy smokers, and those living in crowded industrial areas are usually most sensitive to air pollution.

A. E. Martin[68] has correlated overall daily mortalities and daily mortality due to bronchitis with mean daily black suspended matter measured at seven stations in the London area during the winter of 1958–59. He noted a positive association between levels of the mean daily concentration of sulfur oxides and all causes of death.

A well documented pollution incident occurred in New York City during the 1966 Thanksgiving holidays.[69,70] Some deaths have been attributed to air pollution during this incident. There was also an increase in the number of visits for bronchitis and asthma treatment at seven of the largest New York hospitals during the third day of the episode.

British studies of acute episodes of air pollution show excessive deaths occurring when the smoke levels associated with high levels of SO_2 were over 750 $\mu g/m^3$. Such excessive mortality was also coupled with increased illness. The most sensitive people are older individuals with chronic obstructive pulmonary diseases and those with cardiac problems.

As was pointed out in section 7.6, eye irritation is a major effect of photochemical smog. A survey conducted in southern California showed that more than three-fourths of the population of Los Angeles suffered from eye irritation when the oxidant concentration in ambient air increased to about 200 $\mu g/m^3$ (0.10 ppm). However, such eye irritation resulting from photochemical smog is not only seen in southern California. Many cities like Denver in Colorado, and México, D.F., in México have recorded similar cases of eye irritation due to photochemical smog.

Exposures to high levels of carbon monoxide can increase blood COHb (carboxyhemoglobin) concentration in human subjects. Some researchers[71] have found no decrease in body performance during three-hour exposures to 50–250 ppm CO. They concluded that if CO at these levels has an initial adverse effect adaptive processes must take place during the early phase of exposure, and the compensatory changes override the initial effect of carbon monoxide.

Hydrocarbons per se are not detrimental to human health, however, they have a direct influence on the levels of photochemical oxidants which in turn have adverse health effects.

Many attempts have been made by individual researchers and teams sponsored by health authorities to correlate air pollution levels with mortality and morbidity. Data for respiratory diseases, cancer, and cardiovascular diseases seem to indicate higher mortality and morbidity rates whenever air pollution is high, especially for those over fifty. However, the results of many of these studies have to be looked at with some skepticism since it is very difficult to ascribe a specific death or illness to a pollution episode.

Effects on Animals

Numerous laboratory studies have been carried out to determine the effects of pollution on animals. Research of this nature is aimed at extrapolating the toxicological effects of air pollutants on man. As an example, we can mention the work of Pattle and Burgess[72] who studied the effects of mixtures of SO_2 and smoke on mice and guinea pigs.

The interaction of carcinogens with atmospheric particulates has been studied toxicologically utilizing laboratory animals. Addition of inert particles to carcinogens results in the production of malignant neoplasms in the lungs. Animals have also been used to investigate the effects of ozone. Chemical and biochemical changes have been observed in the heart, liver, and brain of animals following inhalation of ozone. In some experiments adult mice were exposed to 390 $\mu g/m^3$ (0.02 ppm) ozone for five hours daily for three weeks.[29] Structural changes in all membranes and in the nuclei of myocardial muscle fibers were produced. These changes reversed themselves approximately one month after the exposure.

Effects on Plants

Many air pollutants, such as, sulfur dioxide, oxidants (ozone, PAN, etc.) and fluoride compounds can cause injury to vegetation. Much research has been aimed at studying the effects of pollutants on vegetation due to the close relationship of such effects to agricultural production and economic losses by crop damage. These studies are also important because many plants can serve as sensitive indicators of the presence of certain pollutants. One variety of tobacco is used as an effective ozone indicator; pinto bean plants are used to detect peroxyacetyl nitrate (PAN); petunias are used to investigate total oxidants; and gladioli to determine fluoride accumulation. Dalias, petunias, alfalfa, and cotton are excellent detectors of sulfur dioxide.[73]

Ozone affects the surface of the leaves of pinto beans. Such injuries have been reproduced experimentally in the most sensitive species after exposure to 0.03 ppm of ozone for eight hours. PAN causes an under-surface glazing or bronzing of the bean leaf.[74,75] This change can be observed in the laboratory after exposing the plant for five hours to 0.01 ppm of PAN.

Numerous researchers have studied the effect of pollution on tobacco leaves.[76–78] Feder[77] reports that a fifty percent reduction in pollen germination and pollen tube growth occurs in tobacco Bel-W_3 after 5.5 hours of exposure to 0.10 ppm of ozone.

Sulfur dioxide can also cause plant damage.[79] Plants are particularly sensitive to SO_2 during periods of intense sunlight and high relative humidity.

The decline of ponderosa pines in the San Bernardino mountains in California appears to be caused by oxidants originating from the Los Angeles area. The United States Department of Agriculture has recently published a report[80] that extensively describes the injuries that air pollution can cause on trees. Readers are referred to this publication for further details on this subject.

Readers must be cautious in evaluating seemingly accurate studies blaming air pollution for plant damage. Many "natural" agents that produce injury to vegetation can easily be confused with damage caused by air pollution. For example, terminal bleach disease in cereal plants can not only be caused by air pollution but also by excessive water loss associated with hot winds.[81] Symptoms similar to sulfur dioxide injury can also be induced by frosts or by mineral deficiency.

Effects on Materials

Pollutants present in our atmosphere can abrade, corrode, tarnish, soil, erode, crack, and discolor materials of all kinds. For instance, particulate matter can erode buildings, corrode metals, destroy painted surfaces, and degrade clothing.

Sulfur dioxide accelerates the corrosion rate of steel. It also attacks and damages a wide variety of building materials such as limestone and marble. SO_2 can cause discoloration of various fibers such as cotton, nylon, and rayon.

Rubber is particularly sensitive to the presence of ozone. It tends to crack in atmospheres containing 0.01 to 0.02 ppm of ozone especially if the rubber is under strain. Some dyes are susceptible to fading when exposed to ozone. The extent of fading depends on the ozone concentration and on the humidity.

Material losses in the United States due to air pollution are beginning to amount to sizable sums of money which impose an added drain to the American economy.

Effects on Visibility

Atmospheric visibility is reduced mostly as a result of scattering and absorption of visible radiation by air molecules and aerosol particles. Particulates are the major culprits of the reduction in visibility.

To illustrate this effect one can note that in a rural atmosphere where the concentration of particulates is of the order of 30 $\mu g/m^3$, the visibility is about 25 miles. On the other hand, an average urban environment with a concentration of particulates of 100 $\mu g/m^3$ offers a visibility range of 7.5 miles. If the concentration of particulates reaches 200 $\mu g/m^3$ the visibility is reduced to 3.5 miles.

Many authors[82] have studied the effect of pollution on the visual range along a given path.[83] The visual range has been found to depend upon the number of particles per unit volume of atmosphere, the cross-sectional area of the particles, the scattering ratio, and the extinction coefficient per unit of path length.

7.11 POLLUTION LEVELS

In previous sections of this chapter we have discussed the sources of air pollution and the type of contamination resulting from each. Air pollution is primarily important in metropolitan areas although some suburban and rural areas may be affected by pollution in a city close by. Pollutants can easily move and diffuse from the core of the city to the nearby rural environment. The extent of such motion will depend on the wind velocity, temperature, class of pollutants, and topographic conditions.

Urban monitoring stations have shown ambient pollutants in concentrations from 3 to 50 times higher than rural stations. Particulate matter is also present in lesser amounts in rural communities. The change in particulate emissions averages from 102 $\mu g/m^3$ at urban stations to 21 $\mu g/m^3$ at remote rural areas.[84]

The estimated background of CO in the lower atmosphere ranges from 0.2 to 6.01 ppm. Concentrations of this pollutant for the period 1962–1967[33] averaged 14.1 ppm in Chicago, 10.6 ppm in Los Angeles, and 7.9 ppm in Philadelphia.

7.12 POLLUTION ABATEMENT ACTIVITIES

The previous sections have attempted to give the reader an idea of the complexity of the air pollution problem. Equally complex are the abatement activities currently being carried out or planned for the immediate future. These abatement activities are the result of efforts by government units, private industry, research laboratories, universities, and concerned citizens.

The regulatory responsibility for air pollution lies mostly in the hands of the Environmental Protection Agency (EPA) and the various statewide regulatory boards. An important step in the program to regulate air quality throughout the United States was taken on April 30, 1971 when national air quality standards for six major pollutants were finalized (Table 7.10). At a press conference a while later William Ruckelhaus, EPA Administrator, stated in relation to these standards that "if we have erred at all in setting these standards, we have erred on the side of public health."

However, it should be noted that some cities may have problems in meeting these standards unless they radically change urban transportation schemes. Particularly, the cities of Chicago, New York, Philadelphia, Los Angeles, Washington, D.C., and Denver need radically improved transit systems in order to relieve the air pollution problem within the urban areas.

TABLE 7.10.

National air quality standards (Protective of Human Health). (*courtesy of Environmental Science and Technology—vol. 5, June 1971*).

Pollutant	*Level not to exceed*	
SO_2	80 $\mu g/m^3$ (0.03 ppm)	(a)
	365 $\mu g/m^3$ (0.14 ppm)	(b)
Particulate matter	75 $\mu g/m^3$	(c)
	260 $\mu g/m^3$	(b)
Carbon monoxide	10 $\mu g/m^3$ (9 ppm)	(d)
	40 $\mu g/m^3$ (35 ppm)	(e)
Photochemical oxidants	160 $\mu g/m^3$ (0.08 ppm)	(e)
Hydrocarbons	160 $\mu g/m^3$ (0.24 ppm)	(f)
Nitrogen oxides	100 $\mu g/m^3$ (0.05 ppm)	(a)

(a) annual arithmetic mean
(b) maximum 24-hr concentration not to be exceeded more than once a year
(c) annual geometric mean
(d) maximum 8-hr concentration not to be exceeded more than once a year
(e) maximum 1-hr concentration not to be exceeded more than once a year
(f) maximum 3-hr concentration (6–9 am) not to be exceeded more than once a year

TABLE 7.11.

Industrial pollution cleanup expenditures.

Industry	Total investment required (billion dollars)	Millions of dollars 1970 actual	Millions of dollars 1971 estimated	Spending percent change, 1970 versus 1971
Electric utilities	$3.24	$405	$679	+68%
Iron and steel	2.64	206	212	+3%
Petroleum	2.12	337	507	+50%
Paper	1.84	153	321	+110%
Nonferrous metals	1.62	100	152	+52%
Gas utilities	1.04	110	148	+35%
Chemicals	1.00	169	263	+56%
Mining	0.74	115	135	+17%
Machinery	0.69	121	169	+40%
Food and beverages	0.40	84	151	+80%
Commercial	0.32	100	158	+58%
Railroads	0.32	28	28	0
Rubber	0.30	50	42	−16%

Data from McGraw Hill Economics Department; total required by pollution control standards as of January 1, 1971; commercial category includes stores, insurance companies, and banks. (From Business Week, 5/15/71; copyright 1971 McGraw Hill).

Industry is also contributing a large share to the pollution abatement battle. Table 7.11 shows that electric utilities lead the list for overall air and water pollution control expenditures with an estimated $3.24 billion (estimated expenditures of $679 million in 1971, up 68% over 1970). Other big spenders are the iron and steel, petroleum, and paper industries.

Research laboratories and universities are also contributing to air pollution abatement by:

(a) carrying out research to improve air quality, increase effectiveness of combustion processes to reduce stack emissions, produce less-polluting modes of transportation, etc.
(b) training environmental engineers, scientists, and scholars to solve our pollution problems, and
(c) increasing the individual citizen's awareness of environmental problems.

While government and industry can do much to abate air pollution and improve our management of the environment, final responsibility for the cleanup lies in the individual citizen. Modes of urban transportation are going to have to change especially within our large urban communities. While this may cause some inconvenience for the individual, it is the price we need to pay to obtain a clean environment.

One of the most important anti-pollution activities is the research being carried out by the automotive and petroleum industries to improve automotive emission levels. Such research involves work on improved fuel performance, system analysis, use of computer simulators, and actual testing of concept emission vehicles.[85–89]

To improve automotive emission levels work needs to be done both on the car's design and on the fuel used. Work to improve the anti-pollution characteristics of the fuel includes the development of additives to [89]:

(a) suppress formation of oxides of nitrogen,
(b) improve hydrocarbon combustion, and
(c) enhance the atomization of carburetor/induction systems.

Work on new car design to reduce pollution involves[89]:

(a) Design of exhaust control systems that can operate at almost peak efficiency from the instant one turns on the engine.
(b) Optimization of hydrocarbons/carbon monoxide oxidation catalysts and converter systems. One suggested way of achieving this is by using an electrically heated catalytic converter that can heat up to 300°C in thirty seconds (Figure 7.4).
(c) Design of thermal reactors to reduce emissions of CO and NO_x.

The best method to see whether or not certain car-emission objectives have been attained is to subject a concept-emission vehicle to development, durability, exhaust emission, and evaporative emission testing. Many such vehicles have been built and have undergone or are undergoing testing. As an illustration we can refer to a concept emission-free vehicle developed by IIEC (Inter-Industry Emission Control Program) in recent years (Figure 7.5). Taylor and Campau[89] describe the major components of this vehicle which include:

"The limited use of exhaust gas recirculation during part throttle operating modes to minimize NO_x emissions while maintaining acceptable driveability.
An internally-vented carburetor to control carburetor "hot soak" evaporative emissions.
A secondary air pump for the required oxygen for the second converter.
A "Programmed By-pass System" valve to control the flow of the exhaust gases through or around the catalyst converters.
A catalytic converter operating in a reducing atmosphere (excess CO

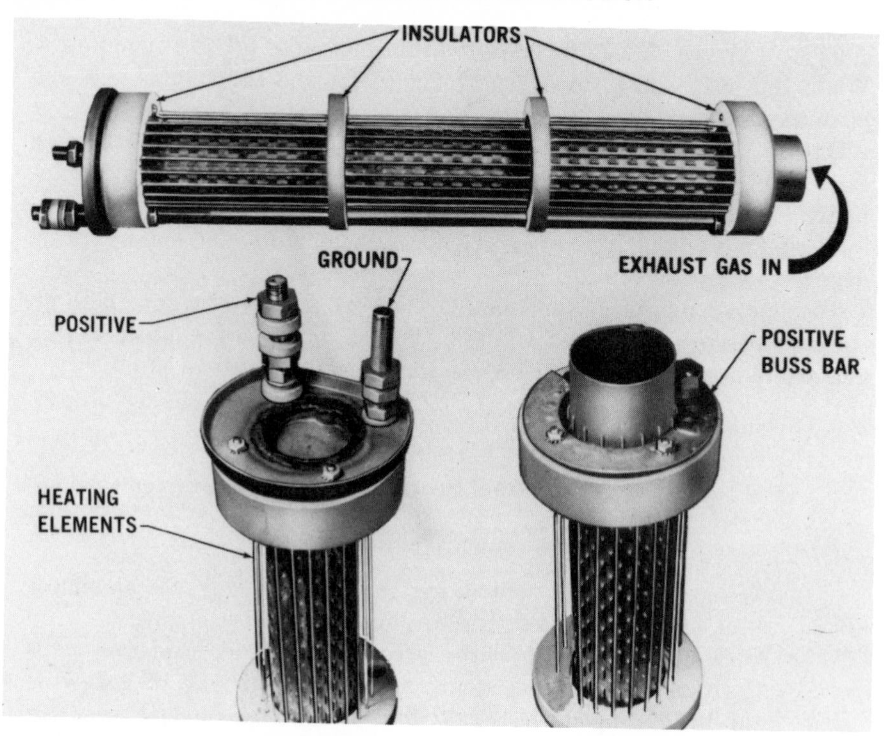

Figure 7.4. Electrically heated hydrocarbon/carbon monoxide catalytic converter (courtesy R. M. Campau).

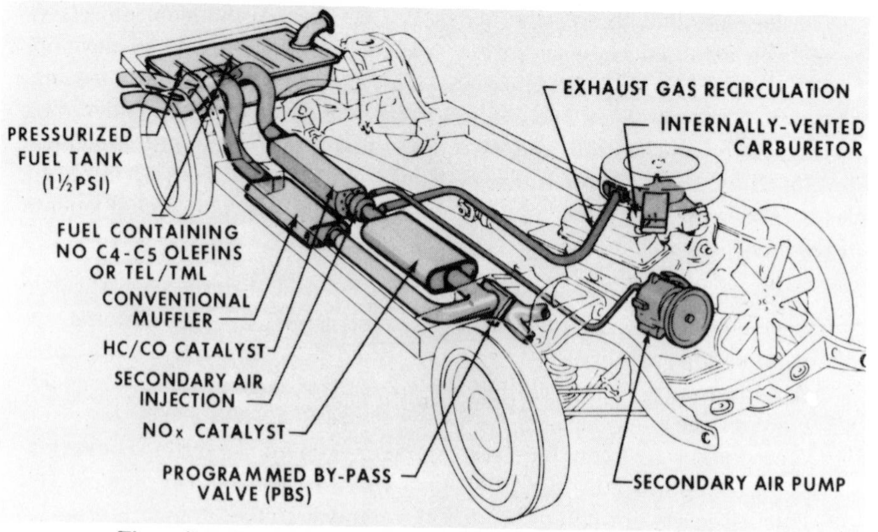

Figure 7.5. A concept emission vehicle (courtesy R. M. Campau).

and H_2) to provide further catalytic reduction of the exhaust NO_x emissions.

A supplemental air injector which injects gases into the system before the exhaust gases pass into the hydrocarbon and carbon monoxide catalyst.

A second catalytic converter which reduces the hydrocarbon and carbon monoxide emissions of the exhaust.

A pressurized gasoline tank [(1½) psig] containing a modified fuel having no C_4–C_5 olefins for evaporative emission reasons and no TEL/TML compounds to extend catalyst life."

In addition to the above-described work to develop new pollution-free cars, much work is in progress to design more efficient cars and automotive products for the coming years.

7.13 REFERENCES

1. Stern, A. C. (ed.) (1968) "Air Pollution", second edition, volumes I and II, Academic Press, New York, N.Y.
2. Leighton, P. A. (1961) "Photochemistry of Air Pollution," Academic Press, New York, N.Y.
3. Calvert, J. C., and Pitts, Jr., J. N. (1966) "Photochemistry," John Wiley and Sons, New York, N.Y.
4. Stanford Research Institute (1961) "Chemical Reactions in the Lower and Upper Atmosphere," Proceedings of an International Symposium, arranged by S.R.I., Interscience Publishers, New York, N.Y.
5. The U.S. Surgeon General (1962) "Motor Vehicles, Air Pollution and Health," U.S. Dept. of Health, Education, and Welfare, Public Health Service Document 489, Washington, D.C.
6. Byres, H. G. (1959) "General Meteorology," third edition, McGraw Hill Book Co., Inc., New York, N.Y.
7. Secretary of H.E.W. (1970) "The Cost of Clean Air. Second Report of the Secretary of H.E.W." H.E.W.–U.S. Government Printing Office Document No. 91-65, Washington, D.C.
8. Nicolet, M. (1960) *The Properties and Constitution of the Upper Atmosphere,* in "Physics of the Upper Atmosphere" edited by J. A. Ratcliffe, Academic Press, New York, N.Y.
9. I.R.E. (1950) *Standards on Wave Propagation. Definition of Terms,* Proceedings of the I.R.E., volume 38, p. 1264.
10. U.S. Navy Weather Research Facility (1961) "The Upper Atmosphere," chart NWRF 26-1161-051, U.S. Naval Air Station, Norfolk, Va.

11. Anonymous (1970) "Nationwide Inventory of Air Pollutant Emissions, 1968," U.S. Dept. of Health, Education, and Welfare, Environmental Health Service, N.A.P.C.A. Publication AP-73, Raleigh, N.C.
12. Anonymous (1969) "Air Quality Criteria for Particulate Matter," U.S. Dept. of Health, Education, and Welfare, Public Health Service, N.A.P.C.A. Publication AP-49, Washington, D.C.
13. Mie, G. (1908) *Beitragezur Optic Truber Modien Speziell Kolloidaler Metallosungen*, Ann Physik, volume 25, pp. 377–445.
14. Robbins, R. C. and Cadle, R. D. (1968) *Kinetics of the Reaction Between Gaseous Ammonia and Sulfuric Acid Droplets in an Aerosol*, Journal of Physical Chemistry, volume 62, pp. 469–471.
15. Goetz, A. and Pueschel, R. (1967) *Basic Mechanisms of Photochemical Aerosol Formation*, Atmos. Environ., volume 1, pp. 287–306.
16. Chass, R. L. and Rose, A. H. (1953) *Discharge from Municipal Incinerators*, preprint presented at the 46th annual meeting of the Air Pollution Control Association.
17. Dept. of HEW (1969) "Air Quality Criteria for Sulfur Oxides," U.S. Dept. of Health, Education, and Welfare, N.A.P.C.A. Publication Doc. AP-50.
18. Johnstone, H. F., and Coughnawr, D. R. (1958) *Absorption of Sulfur Dioxide from Air Oxidation in Drops Containing Dissolved Catalyst*, Ind. Eng. Chem., volume 50, pp. 1169–1172.
19. Gerhard, E. R., and Johnstone, H. F. (1955) *Photochemical Oxidation of Sulfur Dioxide in Air*, Ind. Eng. Chem., volume 47, pp. 972–976.
20. Junge, C. E., and Ryan, T. (1958) *Study of SO_2 Oxidation in Solution and Its Role in Atmospheric Chemistry*, Quarterly Journal of the Royal Meteorol. Soc., volume 84, pp. 46–55.
21. Hall, T. C. (1953) "Photochemical Studies of Nitrogen Dioxide and Sulfur Dioxide," doctoral thesis, University of California at Los Angeles.
22. Dainton, S. F., and Iving, K. J. (1950) *The Photochemical Formation of Sulphinic Acids from Sulphur Dioxide and Hydrocarbons*, Trans. of the Faraday Soc., volume 46, pp. 374–386.
23. Schuck, E. A., Ford, H. W., and Stephens, E. R. (1958) "Air Pollution Effects of Irradiated Automobile Exhaust as Related to Fuel Composition" Rept. No. 26, Air Pollution Foundation, San Marino, California.
24. Anonymous (1970) "Control Techniques for Nitrogen Oxide Emissions from Stationary Sources," U.S. Dept. of Health, Education, and Welfare, P.H.S., C.P.E.H.S., N.A.P.C.A. Publication AP-67.
25. Stephens, E. R. (1961) *The Photochemical Olefin-Nitrogen Oxides Reactions*, in reference 4, pp. 51–64.

26. Altshuller, P. A., Kopczynsky, L. S., Lonneman, W. A., Becker, T. L., and Slater, R. (1967) *Chemical Aspects of the Photooxidation of the Propylene-Nitrogen Oxide Systems,* Environmental Science and Technology, volume 1, pp. 899–914.
27. Tuesday, S. C. (1961) *Photooxidation of Transbutane-2 and Nitric Oxide,* in reference 4, pp. 15–49.
28. Pomonovsky, C. J., Ingels, M. R., and Gordon, J. R. (1966) "Estimation of Smog Effects in the Hydrocarbon-Nitric Oxide System," paper presented at the Air Pollution Control Association National Meeting, June 20-24, San Francisco, California, paper No. 66–42.
29. Anonymous (1970) "Air Quality Criteria for Photochemical Oxidants," U.S. Dept. of Health, Education, and Welfare, P.H.S., C.P.E.H.S., N.A.P.C.A. Publication AP-63, Washington, D.C.
30. McKee, H. (1961) *Atmospheric Ozone in Northern Greenland,* Air Poll. Control Assoc., volume 11, pp. 562–565.
31. Ruckelhaus, W. (1971) *Introductory Statement by Mr. Ruckelhaus at the Hearing on Motor Vehicle Pollution Control,* Environmental News E.P.A. Washington, D.C., May 6.
32. Pressman, J., Arim, L. M., and Warneck, P. (1970) "Mechanisms for the Removal of Carbon Monoxide from the Atmosphere," N.A.P.C.A. Contract G.C.A. TR. 70-66.
33. Anonymous (1970) "Air Quality Criteria for Carbon Monoxide," U.S. Dept. of Health, Education, and Welfare, P.H.S., N.A.P.C.A. Publication AP-62, Washington, D.C.
34. Swinnerton, J. W., Linnenbon, V. J., and Cheek, C. A. (1969) *Distribution of Methane and Carbon Monoxide Between the Atmosphere and Natural Waters,* Environmental Science and Technology, volume 3, pp. 836–838.
35. Robbins, R. C., Borg, K. M., and Robinson, E. (1968) *Carbon Monoxide in the Atmosphere,* Journal Air Poll. Control Assoc., volume 19, pp. 106–110.
36. Anonymous (1970) "Air Quality Criteria for Hydrocarbons," U.S. Dept. of Health, Education, and Welfare, P.H.S., N.A.P.C.A. Publication AP-64, Washington, D.C.
37. Stephens, E. R., and Scott, E. W. (1962) *Relative Reactivity of Various Hydrocarbons in Polluted Air,* Proceedings Amer. Petr. In., volume 42, pp. 665–670.
38. Hanst, L. P. (1971) *Mechanism of Peroxyacetyl Nitrate Formation,* Journal Air Poll. Control Assoc., volume 21, pp. 269–271.
39. Kopczynski, S. L. (1964) *Photooxidation of Alkyl Benzene-Nitrogen Dioxide Mixtures in Air,* J. Air and Water Pollution, volume 8, pp. 107–120.

40. Altshuller, A. P., and Bufalini, J. J. (1965) *Photochemical Aspects of Air Pollution,* Photo-chem. Photobiol., volume 4(2), pp. 99–146.
41. Stern, C. A. (ed.) (1970) "Proceedings of the Symposium on Multiple-Source Urban Diffusion Models," N.A.P.C.A. Publication AP-86, Raleigh, N.C.
42. McCormick, A. R. (1969) *Meteorology and Urban Air Pollution,* W.M.O. Bulletin, pp. 155–165.
43. U.S. Dept. of H.E.W. (1969) "Tall Stacks, Various Atmospheric Phenomena and Related Aspects," U.S. Dept. of Health, Education, and Welfare, Public Health Service, N.A.P.C.A. Publication APTD69-12.
44. McElroy, L. J. (1969) *A Comparative Study of Urban and Rural Dispersion,* J. Appl. Meteorol., volume 8, pp. 19–31.
45. Church, P. E. (1949) *Dilution of Waste Stack Gases in the Atmosphere,* Ind. Eng. Chem., volume 41, pp. 2753–2756.
46. Sutton, O. G. (1953) "Micrometeorology," McGraw Hill Book Co., Inc., New York, N.Y.
47. Eisenbud, M. (1963) "Environmental Radioactivity," McGraw Hill Book Co., Inc., New York, N.Y.
48. Pasquill, F. (1962) "Atmospheric Diffusion," D. Van Nostrand Company, Inc., Princeton, N.J.
49. Strom, H. G. (1962) *Atmospheric Dispersion of Stack Effluents,* in reference 1, pp. 118–198.
50. Stephens, F. R., and Burleson, R. F. (1969) *Distribution of Light Hydrocarbons in Ambient Air,* paper presented at the 62nd annual meeting of the Air Pollution Control Association, New York, June 22–26.
51. Neligan, R. E. (1962) *Hydrocarbons in Los Angeles Atmosphere,* Arch. Environ. Health, volume 5, pp. 581–591.
52. Engel, E. R., Hammer, I. D., Horton, M. J. R., Lane, M., and Plumlee, A. L. (1971) "Environmental Lead and Public Health," United States Environmental Protection Agency Publication AP-90.
53. Tripler, B. A., Jr., and Smithson, G. R., Jr: (1970) *A Review of Air Pollution Problems and Control in the Ceramic Industries,* paper presented at the 72nd annual meeting of the American Ceramic Society, Philadelphia, Pa.
54. Public Health Service (1968) "1968 National Survey of Community Solid Waste Practices," An Interim Report, U.S. Dept. of Health, Education, and Welfare, Public Health Service, Cincinnatti, Ohio, P.H.S. Publication 1867.
55. Jutze, G. A., and Tabor, E. C. (1963) *The Continuous Air Monitoring Program,* Journal Air Poll. Control Assoc., volume 13, pp. 278–280.

56. American Public Health Assoc. (1969) "Guide to the Appraisal and Control of Air Pollution," American Public Health Assoc., New York, N.Y.
57. Staff U.S. Bureau of Mines (1967) "Ringelmann Smoke Chart" (Revision of IC-7718), U.S. Department of the Interior Bureau of Mines IC-8333.
58. Parr, S. W., and Staley, W. D. (1931) *Determination of Sulfur by Means of the Turbidimeter,* Ind. Eng. Chem. Anal. Edition, volume 3, pp. 66–67.
59. Interbranch Chemical Advisory Committee (1965) "Selected Methods for the Measurement of Air Pollutants," U.S. Dept. of Health, Education and Welfare, Public Health Service Publication 999-AP-11.
60. Barnes, H. (1950) *A Modified 2:4 Xylenol Method for Nitrate Estimation,* Analyst, volume 75, p. 388.
61. Farmer, J. R., and Williams, J. D. (1966) "Interstate Air Pollution Study Phase II Project Report-III, Air Quality Measurements," U.S. Dept. of Health, Education, and Welfare, Public Health Service, Cincinnati, Ohio.
62. Byers, D. H., and Saltzman, B. E. (1958) *Determination of Ozone in Air by Neutral and Alkaline Iodide Procedures,* Journal American Ind. Hygiene Assoc., volume 19, pp. 251–257.
63. Sawicki, E., Hauser, T. R., Stanley, T. W., and Elbert, W. (1961) *The 3-Methyl-3-benzothiazolone Hydrazone Test,* Analytical Chemistry, volume 33, p. 93.
64. West, P. W., and Gaeke, G. C. (1956) *Fixation of Sulfur Dioxide as Disulfitomercurate (II), Subsequent Colorimetric Estimation,* Analytical Chemistry, volume 28, p. 1916.
65. Hochheiser, Seymour (1964) "Methods of Measuring and Monitoring Atmospheric Sulfur Dioxide," U.S. Dept. of Health, Education, and Welfare, Public Health Service Publication 999-AP-6, Cincinnati, Ohio.
66. Katz, M. (1952) *The Photoelectric Determination of Atmospheric Sulphur Dioxide by Dilute Starch-Iodine Solutions,* chapter 71 in "Proceedings of the United States Technical Conference in Air Pollution" edited by L. C. McCabe, McGraw Hill Book Co., Inc., New York, N.Y.
67. Thomas, M. D., MacLeod, J. A., Robbins, R. C., Goettelman, R. C. Eldridge, R. W., and Rogers, L. H. (1956) *Automatic Apparatus for Determination of Nitric Oxide and Nitrogen Dioxide in the Atmosphere,* Analytical Chemistry, volume 28, pp. 1810–1816.
68. Martin, A. E. (1964) *Mortality and Morbidity Statistics and Air Pollution,* Proc. Roy. Soc. Med., volume 57, pp. 969–975.
69. Glasser, M., Greenberg, L., and Field, F. (1967) *Mortality and Mor-*

bidity During a Period of High Levels of Air Pollution, New York, November 23-25, 1966, Arch. Environ. Health, volume 15, pp. 674–684.
70. McCarrol, J. R., Cassel, E. G., Walter, E. W., Mountain, J. D., Diamond, J. R., and Mountain, I. R. (1967) *Health and the Urban Environment. V. Air Pollution and Illness in a Normal Urban Population*, Arch. Environ. Health, volume 14, pp. 178–184.
71. Theodore, J., O'Donnell, D. R., and Back, C. K. (1970) *Toxicological Evalation of Carbon Monoxide in Humans and Other Mammalian Species*, in "Lectures in Aerospace Medicine" U.S.A.F. School of Aerospace Medicine, Brooks A.F.B., Aerospace Medical Div. Seventh Series, pp. 194–232.
72. Pattle, R. E., and Burgess, F. (1957) *Toxic Effects of Mixtures of Sulfur Dioxide and Smoke with Air*, Journal Pathol. Bacteriol., volume 73, pp. 411–419.
73. Ibrahim, J. (1970) "Air Pollution Injury to Vegetation," U.S. Dept. of Health, Education, and Welfare, Eivnronmental Health Service, N.A.P.C.A. Publication AP-71, Raleigh, N.C.
74. Middleton, J. T., and Haagen-Smit, A. J. (1961) *The Occurrence, Distribution, and Significance of Photochemical Air Pollution in the United States, Canada, and Mexico*, Journal Air Poll. Control Assoc., volume 11, pp. 129-134.
75. ——— (1964) *Trends in Air Pollution Damage*, Arch. Environ. Health, volume 8, pp. 19–24.
76. Heggestad, H. E., Burleson, F. R., Middleton, J. T., and Darley, E. F. (1964) *Leaf Injury on Tobacco Varieties Resulting From Ozone, Ozonated Hexane-1, and Ambient Air of Metropolitan Areas*, Air and Water Poll., volume 8, pp. 1–10.
77. Feder, W. A. (1968) *Reduction in Tobacco Pollen Germination and Tube Elongation Induced by Low Levels of Ozone*, Science, volume 160, p. 1122.
78. Heggestad, H. E. (1966) *Ozone as Tobacco Toxicant*, Journal Air Pollution Control Assoc., volume 16, pp. 691–694.
79. Hindawi, I. J. (1968) *Injury by Sulfur Dioxide, Hydrogen Fluoride, and Chlorine as Observed and Reflected on Vegetation in the Field*, Journal Air Pollution Control Assoc., volume 18, pp. 307–312.
80. Forest Service (1971) "Air Pollution and Trees," U.S. Dept. of Agriculture, Forest Service Atlanta, Ga.
81. Threshow, M. (1965) *Evaluation of Vegetation Injury as an Air Pollution Criterion*, Journal Air Poll. Control Assoc., volume 15, pp. 266–269.
82. Robinson, E. (1962) *Effects of Air Pollution in Visibility*, in reference 1, pp. 220–254.
83. Huschke, R. E. (1959) "Glossary of Meteorology," Meteorology Soc., Boston, Mass.

84. Ludwig, H. J., Morgan, B. G., and McMullen, B. T. (1969) *Trends in Urban Air Quality*, paper presented at the national fall meeting of the American Geophysical Union, San Francisco, California, Dec. 17.
85. Campau, R. M. (1971) *Low Emission Concept Vehicles*, paper presented at the 36th midyear meeting of the American Petroleum Institute-Division of Refining, San Francisco, California, May 12.
86. Kaneko, Y., Kuroda, H., and Tanaka, K. (1971) *Small Engine-Concept Emission Vehicles*, SAE Preprint # 710296.
87. Schwochert, H. (1969) *Performance of a Catalytic Converter on Non-Leaded Fuel*, SAE Preprint # 690503.
88. Newhall, H. K. (1967) *Control of Nitrogen Oxides by Exhaust Recirculation—A Preliminary Theoretical Study*, SAE Preprint # 670495.
89. Taylor, R. E., and Campau, R. M. (1969) *The IIEC—A Cooperative Research Program for Automotive Emission Control*, paper presented at the 34th midyear meeting of the American Petroleum Institute-Division of Refining, Chicago, Illinois.

Appendix 1

EXERCISES

Exercises are included in this appendix covering all chapters of the book. These problems are included for the convenience of an instructor using this book as a text for a course. The exercises include simple questions, definitions, essay-type questions, descriptive problems, numerical problems, decision-making questions, and requests for library searches. Exercises vary from quite simple ones to rather difficult and thought-provoking ones. Such a variety is aimed at satisfying as many users of the book as possible. All problems included here have been used by the authors at the freshman-sophomore level.

CHAPTER 1. EXERCISES

1. In Chapter 1 it was noted that air can best be classed as a "common property resource." It was further noted that such a resource cannot be managed on the basis of the price system. How would you go about regulating the use of air space by both individuals and corporations? What implications will your management scheme have on an international scale? Can a similar scheme be applied to other "common property resources"?

2. Why is the environment in a dynamic state? What forces are at work?

3. Why should one consider geomorphological degradation as a form of pollution?

4. Why is the problem of overpopulation so intimately related to pollution of the environment? What can we do about overpopulation that may help curb pollution? Discuss various alternatives.

5. Explain what is meant by the principle of superabundance of resources. Discuss the implications of a social system based on this principle.

6. What are some of the factors that need to be considered in order to plan for the optimal utilization of our environmental resources? Discuss and illustrate.

7. What factors must be considered in determining the type of environment that we want future generations to inherit from us?

8. Explain what is meant by multiple sequential land use. Where in your community could you apply this idea? How would you carry it out?

9. How can you reconcile the need for developing new energy sources to satisfy increasing demands with the degradation of the environment as a result of producing energy? How would you regulate the energy industry so as to preserve the environment and at the same time guarantee to the public the availability of energy?

10. What items should be included in a comprehensive national policy in the area of waste disposal?

CHAPTER 2. EXERCISES

1. What phases of environmental science are included under the subject of geomorphology?

2. Explain the meaning of Playfair's Law.

3. Why are land forms dynamic features?

4. Define peneplain.

5. Discuss the implications of the principle of dynamic equilibrium of land forms.

6. What interaction exists between denudation and accumulation processes?

7. Why are adsorption-desorption processes important in studying the products of weathering? How does weathering affect the landscape?

8. Is diastrophism a slow or a fast process? What are the implications of extensive diastrophism in a given area?

9. What is the effect of topography on chemical weathering?

10. How is the erosive work of streams affected by climatic changes?

11. What is the effect of a vegetal cover on soil erosion?

12. Why are cloudburst floods an important factor in soil erosion?

13. What has been the role of man in re-shaping the earth's landscape? What role do other members of the animal kingdom play? Illustrate.

14. How can subsidence create cracks in the earth's surface?

15. It was pointed out in Chapter 2 that excessive withdrawal of underground fluids may lead to subsidence under certain circumstances. What problems may result from excessive injection of fluids into underground strata? Under what conditions would they occur? What could be done to stop them?

16. What advantages are offered by space photography over conventional aerial photography as far as a geomorphological study is concerned?

CHAPTER 3. EXERCISES

1. Compare the importance of the oceans with that of rivers in relation to:

   (a) water supply,
   (b) transportation,
   (c) defense, and
   (d) extractable riches.

2. What are the advantages and disadvantages in using groundwater or surface waters as sources of water supply?
3. Why is the square of the velocity of a river directly proportional to the river's cross sectional area and its slope?
4. Discuss the importance of water movement in preserving the quality of the water.
5. Compare the importance of the unsaturated and saturated regions from a water supply standpoint.
6. How does a permeable soil affect evaporation, infiltration, and runoff?
7. Using dimensional analysis derive equation (3.7-2).
8. How can mineral composition and grain texture affect the intrinsic permeability of a sediment?
9. Consider a gravel mixed with a very fine sand. If the porosity of the gravel is 30% and that of the sand is 35%, calculate what the resulting porosity will be if we mix one cubic foot of sand with 2.5 cubic feet of gravel?
10. What is the minimum porosity that a mixture of gravel of 30% porosity and sand of 40% porosity can have?
11. Consider a confined aquifer 50 feet thick and occupying an area of 20 square miles. The specific yield of this aquifer is 18% and there is an average recharge into it of 100,000 gallons per day. What will be the lowering of the piezometric surface (if any) that would result due to pumping 500,000 gallons per day for 25 years?
12. Why is the storage coefficient low for limestone aquifers?
13. Derive equation (3.7-9) from (3.7-8). What were Jacob's assumptions in obtaining the modified well equation? Under what conditions is equation (3.7-9) applicable?
14. What should be the basic goal of a community's water supply program?
15. Discuss the geographical distribution of surface and subsurface waters in the United States.
16. Describe the type of constituents most commonly present in natural waters.

17. Search your library for information on desalting techniques. On the basis of the information gathered and the discussion on this chapter, evaluate the advantages and disadvantages of desalting techniques.

18. Discuss the most serious water pollution problems that your community has had in the past ten years. Search as many sources of information as possible. Evaluate what was done to abate the pollution and determine whether the pollution incident(s) could have been avoided.

19. Define and/or illustrate the following concepts:

(a) eutrophication,
(b) hydrologic inventory,
(c) water table,
(d) aeration,
(e) runoff,
(f) sublimation,
(g) groundwater reserves,
(h) leaky aquifer,
(i) biological oxygen demand, and
(j) mineral flotation.

20. Consider a hypothetical suburb of a large city (2.5 million inhabitants) in the northeastern United States. The suburb, which we will call *Bellus*, has a population of 1000 inhabitants. However, at present two large industrial research centers are considering the possibility of relocating in Bellus. These two centers will employ approximately 8000 people. Due to such expansion the population of Bellus is supposed to reach 15,000 within five years. The industrial research centers want to obtain their water from the Bellus' government rather than have to get themselves involved in the water supply business. Their water usage is moderate and they will not be discharging much waste. It should be added that there is a large river that runs near Bellus. Water from this river is slightly polluted, however, with adequate treatment it can be used. Not much groundwater exploration has been done in the area, however, it is the opinion of a reputable geologist in the nearby city, that a reasonably good limestone aquifer can be tapped in the area.

You are the head of a local consulting firm and the following questions are put to you by the government of Bellus. You must answer them only on the basis of the above information and making any inferences that may relate to your own hydrologic knowledge.

(a) What additional information would you like to have in order to evaluate the water supply situation in Bellus?
(b) On the basis of the information above what plan for water supply would you favor?

(c) What priorities, if any, would you favor concerning usage of water in the community?
(d) Would you be in favor of some recycling scheme?
(e) Would it be advisable for the research centers to develop their own water supply source or should they try to depend on the municipality for their water?
(f) What plan would you favor concerning treatment of waste-water?

21. Search in your library for information, and write a factual essay in support and one against each of the following statements:

"By the year 2000 the water situation in the world will be extremely critical."
"Industry cannot afford alone to pay for all anti-pollution programs needed to clean-up the environment."
"The environmental battle cannot be won."

CHAPTER 4. EXERCISES

1. Define the halocline.
2. Compare the effects of both salinity and temperature on ocean density.
3. Why is "sigma-tee" used?
4. How does the Coriolis force affect the direction of current flow in the ocean? Explain and illustrate.
5. Describe the food chain of the ocean.
6. How can eutrophication occur in an estuary? How can it be prevented? Consult Chapter 3 for additional information on eutrophication.
7. Discuss the importance and availability of offshore minerals. Describe some techniques that can be used to mine these valuable resources. Indicate what type of ores are more amenable to dredging.
8. How does the ocean "heat engine" control ocean circulation?
9. What benefits would be accomplished by clarifying marine law in regards to aquaculture?
10. Why do we say the ocean is for the most part a "biological desert? What can man do to change this?
11. How does the marine environment affect the distribution of plankton?
12. What could be an efficient way of regulating the future mineral exploitation of offshore areas on a worldwide scale?
13. Describe how the benthos can adapt to temporarily tolerate some environmental changes in their habitat.
14. List the six major constituents of ocean water.
15. What are the short-term effects of dumping solid and liquid wastes

in the open ocean? Cite some examples. What legal steps should be taken to regulate dumping of wastes in international waters?

CHAPTER 5. EXERCISES

1. What elements constitute the major components of the earth's crust?
2. Although copper is much less abundant than aluminum in the earth's crust it has always been in greater demand. Why?
3. What makes metals like gold, silver, and platinum precious? Will copper or iron ever become precious?
4. If you currently were the minister in charge of mineral resources for a medium-sized underdeveloped nation and if your country were endowed with major mineral resources such as oil and uranium what would be your national policy in regard to minerals? Defend all your statements.
5. What will the fuel situation of the U.S. be in the year 2000?
6. What are the characteristics of mineral resources that distinguish them from other natural resources?
7. What are the basic differences between fusion and fission?
8. How can infrared photography be used in mineral exploration?
9. Explain how vegetable matter such as trees, plants, and leaves can be used as guides in mineral exploration.
10. Why is the mining industry regarded as a major polluter of the environment? What are some of the environmental problems created by mining operations?
11. What are the major steps in the recovery of metals from mined ores? What are some of the environmental problems associated with metallurgical processing?
12. You are the manager of a medium-size smelter in the southwestern United States. The smelter is reasonably close to only one industrialized city. If you had to decide between producing sulfuric acid or elemental sulfur from your smelter smoke, which by-product would you select? What factors should be considered in making such a decision?
13. Can bacteria leaching be applied to all types of ore deposits?
14. What are some of the most important advantages offered by chemical mining? What environmental damage can result as a consequence of this type of operation?

CHAPTER 6. EXERCISES

1. How and why has the disposal of solid waste become a significant problem at the present time?

2. What types of materials can best be recycled?

3. Compare the problems encountered by a municipality searching for a suitable sanitary landfill location with the problems involved in locating a central incinerator?

4. What materials are acceptable for sanitary landfills but not for incineration?

5. How does the geology and hydrology in the vicinity of a municipality affect the selection of a waste disposal method?

6. What conditions must be maintained in an incinerator or sanitary landfill to make it environmentally acceptable?

7. Compare the responsibilities of industry and municipality as far as refuse disposal is concerned.

8. Discuss some possible solutions to the "junk yard" problem?

9. What is the role of recycled metals in international trade today? Can they ever be expected to play a predominant role in our modern economy?

CHAPTER 7. EXERCISES

1. What is the volume of one mole of dry air at standard temperature and pressure (STP) conditions? Use $R = 8.2 \times 10^{-2}$ liter-atmosphere $(°K)^{-1}$ $mole^{-1}$, $T = 273°K$, $P = 1$ atmosphere.

2. What is the volume of oxygen, nitrogen, argon, and carbon dioxide present in one mole of dry air?

3. The average diameter of airborne contaminant X is 50 microns. What is the settling velocity of such particles in still air? Use the following values: density of $X = 11$ g/cc, density of air $= 1.2 \times 10^{-3}$ g/cc, and viscosity of air $= 1.82 \times 10^{-4}$ poises.

4. The 48-contiguous states of the U.S. occupy an area of approximately 3×10^6 square miles. Over this area about 30 million tons of particulates are emitted every year. How many grams of particulates are emitted per day per square centimeter of area? If the particulates are only emitted over 1% of the land surface how many grams are emitted per day per square centimeter of area?

5. A power plant burns 12 tons of fuel oil per hour. The oil being used contains 2% (by weight) sulfur. How much sulfur does such a plant burn in a year?

6. Why are mobile sources of air pollution difficult to control?

7. Describe briefly the major air pollutants in the United States.

8. How can one account for the observed concentrations of sulfate ion in rain, fog, and smog?

9. How can one account for the observed concentrations of ozone in polluted atmospheres?

10. How can the ocean be a source of carbon monoxide?

11. What is the difference between alicyclic and aromatic hydrocarbons?

12. What is photochemical smog?

13. What is a superadiabatic condition? What type of atmospheric movement is favored by such a condition?

14. Describe and illustrate the five basic types of plumes.

15. What are the effects of various methods of solid waste disposal on air quality? (Consult Chapter 6 for additional material on this subject).

16. Suppose you are the head of the Air Pollution Department of the City of Topel. The town's population is 300,000 and there are 100,000 cars. The city is located in a river valley with an area of 200 square miles. The river cuts through the northern part of town. The city has medium to light industry (mostly small-scale manufacturing). The power plant is located on the northern edge of town by the river side.

(a) What criteria would you use in setting up an air quality monitoring program for this city?

(b) What pollutants would you measure to determine the air pollution levels of Topel?

(c) What preventive steps would you take if a city-wide pollution incident threatens?

Appendix 2

CONVERSION FACTORS

Commonly used units

to convert from	into	multiply by
acres	square feet	43,560.0
acres	square meters	4,047.0
acres	square miles	1.56×10^{-3}
acre-feet	gallons	3.26×10^{5}
angstrom (Å)	meter	1×10^{-10}
angstrom (Å)	micron	1×10^{-4}
atmospheres	pounds per in^2	14.70
atmospheres	cms of mercury	76
barrels (U.S. liq.)	gallons	31.5
barrels of oil	gallons of oil	42.0
bars	pounds per in^2	14.50
BTU	ergs	1.06×10^{10}
centimeters	feet	3.28×10^{-2}
centimeters	inches	0.394
centimeters of mercury	atmospheres	0.0132
centimeters/second	knots	0.194
centimeters/second	miles/hour	0.0224
cubic centimeters	cubic feet	3.53×10^{-5}
cubic centimeters	cubic inches	0.061
cubic feet	cubic centimeters	28,320.0
cubic feet	cubic inches	1,728
cubic feet	cubic meters	0.02832
cubic feet/second (cfs)	gallons/minute (gpm)	448.83
cubic inches	gallons	4.33×10^{-3}
darcy	$centimeters^2$	1×10^{-8}
dynes/$centimeter^2$	bars	1×10^{-6}
ergs	BTU	9.48×10^{-11}
fathoms	feet	6
fathoms	meters	1.83

feet	meters	0.305
feet of water (head)	pounds per in^2	0.434
feet/second	knots	0.6
feet/second	miles/hour	0.68
gallons	cubic centimeters	3,785
gallons	cubic feet	0.134
gallons (U.S)	gallons (Imp.)	0.83
hours	days	4.17×10^{-2}
inches	centimeters	2.54
knots	statute miles	1.15
liters	gallons (U.S. liq.)	0.264
meters	feet	3.281
meters	yards	1.094
meters/second	miles/hour	2.24
miles (naut)	kilometers	1.853
miles (stat)	kilometers	1.609
miles/hour	feet/second	1.47
millimicrons	meters	1×10^{-9}
milligrams/liter	parts per million	1.0
million gallons/day	cubic feet/second	1.547
poise	gram/centimeter.sec	1.00
square centimeters	square inches	0.155
square feet	square meters	0.093
square kilometers	square miles	0.386
square meters	square feet	10.76
tons (long)	kilograms	1,016
tons (long)	pounds	2,240
tons (long)	tons (short)	1.12
tons (metric)	kilograms	1,000
yards	centimeters	91.44

Conversion of parts per million (ppm) to equivalents per million (epm)

Ion	Multiply by	Ion	Multiply by
Aluminum + 3	0.11119	Iron + 3	0.05372
Bicarbonate	0.01639	Magnesium + 2	0.08224
Calcium + 2	0.04990	Manganese + 2	0.03640
Carbonate	0.03333	Potassium + 1	0.02558
Chloride − 1	0.02820	Sodium + 1	0.04350
Iron + 2	0.03581	Sulfate	0.02082

Partial listing of the elements

(Based on the atomic mass of $C^{12} = 12.000$)

Element	Symbol	Atomic number	Atomic mass
Aluminum	Al	13	26.98
Antimony	Sb	51	121.75
Arsenic	As	33	74.92
Barium	Ba	56	137.34
Beryllium	Be	4	9.012
Bismuth	Bi	83	208.98
Boron	B	5	10.81
Bromine	Br	35	79.91
Calcium	Ca	20	40.08
Carbon	C	6	12.011
Chlorine	Cl	17	35.45
Chromium	Cr	24	52.00
Copper	Cu	29	63.54
Fluorine	F	9	19.00
Gold	Au	79	196.97
Helium	He	2	4.00
Hydrogen	H	1	1.01
Iodine	I	53	126.90
Iron	Fe	26	55.85
Lead	Pb	82	207.19
Lithium	Li	3	6.939
Magnesium	Mg	12	24.31
Manganese	Mn	25	54.94
Mercury	Hg	80	200.59
Molybdenum	Mo	42	95.94
Nickel	Ni	28	58.71
Nitrogen	N	7	14.0067
Oxygen	O	8	15.999
Phosphorus	P	15	30.97
Platinum	Pt	78	195.09
Potassium	K	19	39.102
Silicon	Si	14	28.09
Silver	Ag	47	107.87
Sodium	Na	11	22.99
Sulfur	S	16	32.064
Tin	Sn	50	118.69
Uranium	U	92	238.03

Vanadium	V	23	50.94
Zinc	Zn	30	65.37

Temperature conversion formulas

°C = degree centigrade; °K = degree kelvin; °F = degree fahrenheit
Temp °K = Temp °C + 273.15
Temp °F = Temp °C × 1.80 + 32°F

AUTHOR INDEX

SUBJECT INDEX